NUREG-1521

I0489268

Technical Review of Risk-Informed, Performance-Based Methods for Nuclear Power Plant Fire Protection Analyses

Draft Report for Comment

Manuscript Completed: March 1998
Date Published: July 1998

M. Dey, M. A. Azarm,* R. Travis,* G. Martinez–Guridi*, R. Levine**

Division of Systems Technology
Office of Nuclear Regulatory Research
U.S. Nuclear Regulatory Commission
Washington, DC 20555–0001

*Brookhaven National Laboratory
 Upton, NY 11973

**Formerly With
 National Institute of Standards and Technology
 Building and Fire Research Laboratory
 Gaithersburg, MD 20899

COMMENTS ON DRAFT REPORT

Any interested party may submit comments on this report for consideration by the NRC staff. Comments may be accompanied by additional relevant information or supporting data. Please specify the report number, draft NUREG-1521, in your comments, and send them by November 30, 1998 to:

Chief, Rules Review and Directive Branch
Office of Administration
Mail Stop T-6 D59
U.S. Nuclear Regulatory Commission
Washington, DC 20555-0001

You may also provide comments at the NRC Web Site, http://www.nrc.gov. See the link under "Technical Reports in the NUREG Series" on the "Reference Library" page. Instructions for sending comments electronically are included with the document, NUREG-1521, at the web site.

For any questions about the material in this report, please contact:

Dr. M. K. Dey
Mail Stop T-9F31
U. S. Nuclear Regulatory Commission
Washington, D.C. 20555

Phone: 301-415-6443
E-mail: mkd@nrc.gov

ABSTRACT

The Nuclear Regulatory Commission (NRC) has instituted an initiative for regulatory improvement to focus licensee and NRC resources on risk-significant activities, and to decrease the prescriptiveness of its regulations through performance-based methods that allow licensees increased flexibility in implementing NRC regulations. The NRC has identified risk-informed methods utilizing insights from probabilistic risk analysis (PRA) as a major tool for achieving its goal for regulatory focus. The issue of fire protection requirements has been identified as a regulatory area in which NRC will pursue regulatory improvement. This report presents a technical review and analysis of risk-informed, performance-based methods that are alternatives to those in current prescriptive fire protection requirements or guidance that could allow cost-effective methods for implementing safety objectives, focusing licensee efforts, and achieving greater efficiency in the use of resources for plant safety. A technical analysis of the usefulness of the results and insights derived from these methods (including accounting for the uncertainties in the results) in improving regulatory decisionmaking is presented.

CONTENTS

FIGURES

TABLES

EXECUTIVE SUMMARY

In SECY-94-090, the staff formulated the framework to institutionalize a *Continuing Program for Regulatory Improvement*. This framework was approved by the Commission and the program was initiated in May 1994. The main objective of this program is to improve regulatory efficiency by providing flexibility to licensees for implementing safety objectives in a cost-effective manner, and to use risk information and insights where appropriate to focus NRC and licensee activities in risk-significant areas of its requirements. Furthermore, in COMSECY-96-061, dated April 15, 1997, "Risk-Informed, Performance-Based Regulation" (DSI-12), the Commission recognized that, in order to accomplish the principal mission of the NRC in an efficient and cost-effective manner, it will in the future have a regulatory focus on those licensee activities that pose the greatest risk to the public. In this document, the Commission reiterated its statement in the PRA Policy Statement that the use of PRA technology should be increased in all regulatory matters to the extent supported by the state of the art in PRA methods and data, and in a manner that complements the NRC's deterministic approach and supports the NRC's traditional defense-in-depth philosophy. NRC's requirements for fire protection have been identified as an area in which NRC intends to pursue regulatory improvement toward meeting the above-stated objectives. The intent of this study is to report on a technical review of risk-informed, performance-based methods for fire protection analyses that have become available since the issuance of NRC fire protection requirements and that have the potential to improve the regulatory system by providing additional insights beyond those provided by current prescriptive methods, and organizing a systematic process for evaluating fire protection issues.

The experience with NRC requirements was reviewed to identify opportunities for the application of risk-informed, performance-based methods, while the availability of these methods was determined in a parallel review. The results of these reviews were used to conduct trial applications of risk-informed, performance-based methods to selected areas of requirements for fire protection contained in Appendix R of 10 CFR Part 50. These applications, or case studies, assess the potential for risk-informed, performance-based methods to provide additional insights that would improve regulatory decisionmaking—in evaluating potential alternate means of implementing NRC fire protection safety objectives while accounting for uncertainties in the results of these methods.

This document presents a technical review of information relevant and useful to the process for regulatory improvement. This document is not intended to support any regulatory action by the NRC staff. It is specifically noted that the applications, or case studies, of risk-informed, performance-based methods presented in this report examine and illustrate the potential benefits of such methods for providing new technical information, a more systematic process for judging the acceptability of alternative approaches to prescriptive compliance, and new or improved insights of the risk significance of key event scenarios, including operator actions. The weaknesses and constraints of these applications will need to be further defined, and guidance developed before the implementation of these methods in the regulatory process.

Experience With NRC Requirements

A comprehensive analysis of experience with NRC requirements was conducted through a review of exemptions granted to Appendix R. The technical bases for granting the exemptions and areas in which risk-informed, performance-based methods were or could have been used to provide the basis for the request for approval or granting of the exemptions were identified. The following conclusions are drawn on the basis of review of exemptions to Appendix R granted by the staff:

- The justifications provided by licensees for the request for exemptions, and the technical bases used by the staff for granting the exemptions, were primarily qualitative analyses of combustible loading and effect based on engineering judgment; in a few cases, quantitative analyses using fire models were submitted by licensees as part of the justifications for the exemptions.

- Qualitative analyses and arguments similar to those in recovery models in PRA Human Reliability Analysis (HRA) were used in several submittals for exemptions; however, quantitative PRA or HRA analyses were not submitted at that time.

- Most of the exemptions are in technical areas amenable to the use of risk-informed, performance-based methods that have been developed since the issuance of Appendix R and exemptions granted to that regulation, e.g., fire PRA, including HRA analysis, and modeling the dynamics of fire effects.

Alternate Methods Developed Since Issuance of Appendix R

In parallel to the review of experience with current requirements summarized above, fire PRA and modeling methods that have been developed and used by the NRC and the U.S. nuclear industry for conducting PRA studies, and by licensees for conducting individual plant examinations for external events (IPEEEs) in response to NRC Generic Letter 88-20, Supplement 4, were reviewed. The results of PRAs and the IPEEEs are currently not used to support regulatory decisionmaking for the implementation of NRC fire protection regulation, but have been limited thus far to examine if specific vulnerabilities to fires exist in plants. Since Appendix R was issued in 1980, the probabilistic risk assessment methodology has been developed and used over the last 15 years by the NRC and the U.S. nuclear industry to

(1) determine plant risk from fire events as part of general assessments of the total risk profile from plant operations, and

(2) identify vulnerabilities to fire events and implement cost-effective safety improvements to either eliminate or reduce the

impact of these fire vulnerabilities.

A review of 12 PRA studies conducted by the NRC, EPRI, and nuclear utilities to assess plant risk, including risk from fire events, yielded the following observations:

- Given the same plant configuration and parameters, the absolute results of fire PRAs vary significantly because of the data, methods, and assumptions used (particularly between those sponsored by NRC and EPRI)

- Given the same data, methods, and assumptions, the major differences in estimated fire CDF can be explained by plant-specific system design and the embedded level of redundancies in safety functions.

- Most studies indicate that the majority (in some cases as much as 90 percent) of the risk from fires in nuclear power plants comes generally from three or four fire areas, such as the control room, cable spreading room, and the switchgear room.

- Fire protection analysis using PRA differs in many respects to analysis per NRC requirements in Appendix R. For example, even though most fire PRAs have identified fires in the control room and the cable spreading room as significant contributors to core-melt probability, a coincident loss of offsite power is not included in the scenarios. This is quite different from the regulation in Appendix R, which requires an assumption that offsite power is lost coincident with a fire in the control room. The significance of a control room fire as modeled in PRAs is usually attributable to scenarios other than the loss of offsite power (e.g., a control room fire in a PWR may, among other things, cause the power-operated relief valves (PORVs) to open spuriously).

The review identified various uncertainty issues that have been stated to be associated with fire PRA and modeling. A number of different areas of a fire protection program can be analyzed without the need for fire modeling (e.g., fire protection equipment surveillance and maintenance test intervals). For these cases, the issue of uncertainty can be formally addressed and incorporated in the decisionmaking process. In other cases in which evaluation of the issues necessitates the use of fire modeling, the portion of fire modeling that predicts the fire heat-release rate can be differentiated from the portion that predicts the thermal environment. Larger uncertainty ranges are associated with the predicted heat-release rate than with the thermal environment. The heat-release rate is the driving force for the plume mass flow rate, the ceiling jet temperature, and finally, the hot layer temperature that is driven by energy balance. The fire heat-release rate is dependent on the initial fire size, the growth of fire by propagation and ignition of additional combustibles, and the heat-release rate from these additional combustibles. In any case, the heat-release rate of the fire source, knowing the current state of the art, may be estimated conservatively by using simplified engineering evaluation, subjective judgment, and extrapolation of actual fire events or fire tests.

Finally, a preliminary conclusion has been reached by the NRC staff that the fire PRA and FIVE methods have been successfully used to achieve the objectives of the IPEEE regulatory program to identify plant vulnerabilities to fire events and implement cost-effective safety improvements to either eliminate or reduce the impact of these fire vulnerabilities. The fire IPEEE conducted by the Quad Cities nuclear power station has been cited by the NRC staff as an example of the success of the IPEEE program and use of fire PRA and/or the FIVE methods to identify vulnerabilities not addressed by Appendix R.

Developments and Practices Outside NRC and U.S. Nuclear Industry

Developments and practices outside NRC and the U.S. nuclear industry were also reviewed. The Institute of Protection and Nuclear Safety (IPSN) of the French Atomic Energy Commission (CEA) and the utility Electricité de France have considerable efforts underway for developing and utilizing fire PRAs supported by fire computer codes. The goal of their program is to advance the state of the art of fire models for nuclear plant applications beyond the current state. They have concluded that this tool provides useful information for safety assessments to supplement engineering judgment on which reactor design and fire protection provisions are based. The French program includes research work for fire code development and validation with tests, and application of the developed fire computer code in their fire PRA studies initiated in 1993. They intend to use fire PRAs to identify the most significant locations where vulnerabilities exist and to support the necessary analysis within the framework of the periodic safety assessments conducted every ten years in France for each plant.

The review of developments in the U.S. and foreign building industries revealed a notable move toward the use of performance-based design methods, and to a limited extent risk analysis, to replace current prescriptive requirements. Among the benefits identified are designs to achieve fire safety that are better and less expensive than those achieved with prescriptive code provisions. Although the main goal of fire protection for commercial buildings, that is, life safety, is different from that for nuclear power plants, several features of the fire models and computer codes being used in the building industry that are essential for applications in nuclear power plants are similar. Also, other important goals in building fire safety — the assessments of the fire endurance of walls and floors to determine fire fighting capability, and spread of fire to nearby structures—are applicable to nuclear power plants. Recognizing the benefits of performance-based methods, several countries (New Zealand, Australia, Canada, and U.K.) have modified their building fire laws and regulations to make this transition to performance-based regulation. Australia and Canada are pursuing the use of risk analysis in conjunction with performance-based methods for building fire protection design. More recently, the National Fire Protection Association in the U.S. has also initiated development of performance-based standards.

Since the early 1980s, notable developments have been made for fire safety engineering analysis for building safety using fire models, particularly in the U.S., U.K., and Japan. A number of computer codes have been developed and are currently being used for building fire protection analysis. Recently, an international collaborative effort involving several countries has been initiated to validate fire computer codes being used in the different countries. Several international conferences are now held annually to present and share results, and experiences. Other than efforts in France, a similar level of international activity for developing the capability for performance-based analysis for nuclear power plant fire protection is not evident. One collaborative effort between U.S. and French utilities to compare fire computer codes is noted.

Applications of Risk-Informed, Performance-Based Methods

This review explored and categorized a variety of applications of risk-informed, performance-based methods for protection analyses.

The first general category of methods is those that would support performance-based approaches, but are not necessarily risk-informed, i.e., these methods will support implementation of less-prescriptive safety objectives, but do not directly analyze or utilize risk information.

The second general category of methods is those that would support performance-based and more risk-informed approaches, i.e., these methods will support implementation of less-prescriptive performance criteria, and they analyze or utilize risk information. Based on the review of exemptions to Appendix R and determination of areas that are amenable to risk-informed, performance-based methods, the following categories and applications were developed and chosen to examine the benefits of applying these methods:

A. Performance-Based Analyses

- "Engineering Tools" for Evaluating Fire Dynamics—Bounding Analyses of Combustible Fire Loads

- Reliability Methods

  - Establishing Surveillance Intervals Based on Performance and Reliability

  - Optimizing Test Duration for Appendix R Emergency Lighting

  - Considerations for the Use of Portable Lights for Outdoor Activities

- Fire Computer Codes Based on Zone Models —Analysis of Safe Separation Distance

B. Risk-Informed, Performance-Based Analyses

- Use of Risk Insights in a Qualitative Manner Evaluating Need for Emergency Lighting

- Event Tree Modeling and Delta-CDF Quantifications

  - Analysis of the 72-Hour Criterion To Reach Cold Shutdown

  - Evaluation of Loss-of-Offsite-Power Assumption for Alternative or Dedicated Shutdown Capability

"Engineering tools" based on the principles of thermodynamics, fluid mechanics, heat transfer and combustion have now become more available and can be useful for analysis of unwanted fire growth and spread (fire dynamics). These analyses can be mostly conducted by hand without a computer program, or sometimes with simple computer routines of fire correlations.

"Engineering tools" for certain configurations are available for calculating an equivalent fire severity, adiabatic flame temperature of the fuel in comparison to the damage temperature of the target, fire spread rate, pre-flashover upper layer gas temperature, vent flows, heat release rate needed for flashover, ventilation limited burning, and post-flashover upper layer gas temperature.

With the formulation of appropriate guidance, these tools can be used in a gross and conservative manner to evaluate the adequacy of deviations from prescriptive requirements for configurations with low fire loading, or to establish the basis for fire barrier ratings, safe separation distance, and need for fire detectors and suppression systems in protecting one train for safe shutdown. Since these tools generally employ bounding calculations, results will be conservative but can provide useful information to indicate areas where fire protection features have been overemphasized (or underemphasized).

In cases in which hand calculations cannot be conducted to provide useful results, fire computer codes can be used for more detailed calculations to support an assessment of the fire hazard and predict fire protection system response. These computer codes are based on plume correlations, ceiling jet phenomena, and hot and cold layer development and can predict the temperature of targets exposed to fires, detector and suppression system actuations, and smoke level and transport during fires in certain specific configurations. As with any model or computer code, it is essential to understand the bounds of the configuration and parameters within which these computer codes are valid in order to use the results for developing credible conclusions.

Several reliability-based (based on operating data) methods are available now and are being used in other areas of NRC requirements. For example, NRC requirements in Appendix J of 10 CFR Part 50 (60 FR 49495) allow licensees an option to formulate a performance-based program for containment leakage testing. Such approaches can be used to determine an optimal and adequate maintenance and surveillance test interval for fire protection detection and suppression systems. Reliability analyses can also be used to provide insights on the important parameters to be considered in optimizing the test duration for emergency lighting, and the approximate change in reliability as a function of test duration.

The results of PRAs and other IPEEE analyses, including human recovery modeling, and other more limited analysis, are now available and can be used in a qualitative manner to provide risk insights regarding the impact of alternate approaches. An example is the use of fire PRA results, including human recovery modeling, to

develop the basis for the plant emergency lighting program in lieu of prescriptive requirements (e.g., 8 hours' duration for all plant areas containing safe-shutdown equipment). Risk-significant accident sequences, e.g.; for fire-induced station blackout, can be examined to determine the need for emergency lighting. In some cases, lighting may be required for more than 8 hours.

Fire PRA and other methodologies have inherent in them screening processes that can progressively distinguish between and identify high- and low-risk fire areas. The screening methods employed in fire PRAs and other methods can be used toward formulating a risk-graded fire protection program by identifying and focusing on critical fire areas. Categories, or grades, can be established for currently identified fire areas in plants. A higher level of fire protection could then be extended to fire areas that contribute significantly to plant fire risk. This approach would be in contrast to prescriptive requirements that specify that all structures, systems, and components (SSCs) of one shutdown train be protected from fires by the same measures regardless of the extent of vulnerability of those SSCs to a fire or impact on plant risk if they are damaged.

PRA operator recovery models and delta-CDF calculations are also available now and can be used to supplement the information used to determine the adequacy of alternate approaches. Regulatory guides currently being finalized for implementing specific changes to a plant's licensing basis allows the use of delta-CDF as an indicator of the acceptability of implementing specific changes. Fire PRA methods can be used to calculate the change in core-damage frequency (delta-CDF) for alternate approaches to fire protection, including for evaluating the role of operators for recovery actions. These methods are useful for evaluating the extent to which repairs are appropriate to maintain one train of systems to achieve and maintain shutdown conditions, and the use of non-standard systems for shutdown. The methods can also be used to evaluate and compare alternate means of providing fire protection (by combining separation, fire barriers, and detection and suppression) to safe-shutdown systems.

Application Cost Benefits

Implementation of alternate approaches for fire protection programs has the potential to provide opportunities for cost optimization. For operating reactors, opportunities are limited in areas in which fire protection programs have already been established and recurring maintenance is not necessary. However, if deficiencies are identified as a result of inspections or self-assessments, the one-time savings could be significant. There is a potential for cost reduction in areas in which recurring activities are required, e.g., for surveillance. These costs can be significant when considered over the life of the plant.

Concluding Remarks

The report presents some potential areas of fire protection requirements that are amenable to currently available risk-informed, performance-based methods, and illustrates the manner in which applications could be made. The benefits of these methods are judged to be that they could provide new or improved insights for fire protection analyses, and a more systematic process to judge the acceptability of alternative approaches. These benefits have the potential to improve decisionmaking and increase flexibility in the current regulatory structure. A comprehensive list of applications, further definition of specific weaknesses and contraints for these applications, and guidance on their use will need to be developed prior to implementing these approaches in the regulatory system.

ACKNOWLEDGMENTS

The NRC is grateful to the Institute of Protection and Nuclear Safety of the French Atomic Energy Commission (CEA) and Electricité de France for the information on their initiatives in technical areas discussed in the report. The authors wish to acknowledge members of the NRC staff for their comments and contributions to this report, and to Rayleona Sanders for editing this report.

ACRONYMS AND INITIALISMS

ADS	automatic depressurization system	ERL	expected risk to life
AFW	auxiliary feedwater	ESGR	emergency switchgear room
AHU	air-handling unit	ESW	emergency service water
ANSI	American National Standards Institute		
AOV	air-operated valve	FCE	fire-cost expectation
APCSB	Auxiliary and Power Conversion Systems Branch	FHAR	fire hazards analysis report
		FIVE	fire-induced vulnerability evaluation
ASHRAE	American Society of Heating, Refrigeration, and Air Conditioning Engineers	FPETOOL	Fire Protection Emergency Tools
		FRA	fire risk assessment
		FSAR	final safety analysis report
ASTM	American Society for Testing and Materials	FSES	fire safety evaluation system
		GL	generic letter
BNL	Brookhaven National Laboratory	GSA	General Services Administration
BRI	Building Research Institute (Japan)	HPCI	high-pressure coolant injection
BRP	Big Rock Point	HPCS	high-pressure core spray
BSI	British Standards Institution	HPI	high-pressure injection
BTP	branch technical position	HRR	heat-release rate
B&W	Babcock & Wilcox	HVAC	heating, ventilation, and air conditioning
BWR	boiling-water reactor		
CCW	component cooling water	IAEA	International Atomic Energy Agency
CDF	core-damage frequency		
CEA	Atomic Energy Commission (France)	IEEE	Institute of Electrical and Electronics Engineers
CFAST	Consolidated Model of Fire Growth and Smoke Transport	IP2	Indian Point Unit 2
		IPE	individual plant examination
CFD	computational fluid dynamics	IPEEE	individual plant examination for external events
CFR	*Code of Federal Regulations*		
CHR	containment heat removal	IPSN	Institute of Protection and Nuclear Safety (France)
CIB	International Council for Building Research and Development	IRRAS	Integrated Reliability and Risk Analysis System
CNRS	Centre National de la Recherche Scientifique	ISO	International Organization for Standardization
COMPBRN	Fire Hazard Model for Risk Analysis		
CRD	control rod drive	LER	licensee event report
CS	containment spray	LES	Large Eddy Simulation
		LOCA	loss-of-coolant accident
DCPP	Diablo Canyon Power Plant	LOR	level of resolution
DG	diesel generator	LOSP	loss of normal ac offsite power
DPC	Duke Power Company	LPCI	low-pressure coolant injection
DSIN	Directorate for the Safety of Nuclear Installations (France)	LPCS	low-pressure core spray
		LPI	low-pressure injection
EdF	Electricité de France	LWR	light-water reactor
EPRI	Electric Power Research Institute		

MCC	motor control center	RCP	reactor coolant pump
MCR	main control room	RCZ	radiological control zone
MCS	minimal cutset	RHR	residual heat removal
MITI and	Ministry of International Trade and Industry (Japan)	RMIEP	Risk Methods Integration and Evaluation Program
MOC	Ministry of Construction (Japan)	RPV	reactor pressure vessel
MSIV	main steam isolation valve	RRG	Regulatory Review Group
		RY	reactor-year
NASA	National Aeronautical and Space Administration	SBO	station blackout
		SDC	shutdown cooling
NFPA	National Fire Protection Association	SER	safety evaluation report
		SFPE	Society of Fire Protection Engineers
NIST	National Institute of Standards and Technology	SI	statutory instrument
NMP2	Nine Mile Point Nuclear Station Unit 2	SINTEF	Stiftelsen for Industriell og Teknisk Forskning (Norway)
NPP	nuclear power plant	SNL	Sandia National Laboratories
NRC	Nuclear Regulatory Commission	SPC	suppression pool cooling
NRCC	National Research Council of Canada	SRP	Standard Review Plan
		SRV	safety relief valve
PC	personal computer	SSCs	structures, systems, and components
PCS	power conversion system	SSD	safe shutdown
PDR	public document room	STA	Science and Technology Agency (Japan)
PG&E	Pacific Gas & Electric Co.	STP	South Texas Project
PORV	power-operated relief valve		
PRA	probabilistic risk assessment	TBCW	turbine building cooling water
PVC	polyvinyl chloride	TSD	technical support document
PWR	pressurized-water reactor	UK	United Kingdom
		UL	Underwriters Laboratory
QRA	quantitative risk analysis		
		V/I	value impact
RAM	risk assessment model		
RBCW	reactor building cooling water		
RCIC	reactor core isolation cooling		

1 INTRODUCTION

As part of the regulatory improvement program it established in 1994, the NRC is reviewing current regulations in an effort to improve regulatory focus and cost-effectiveness of implementing regulatory safety objectives. Reactor fire protection has been identified as one of several areas in which the NRC is pursuing regulatory improvement.

The consideration of risk in regulatory decision-making has long been part of NRC's policy and practice. Initially, these considerations were more qualitative and were based on risk insights. The early regulations were more prescriptive and relied on good practices and accepted deterministic standards rather than on quantitative models and risk-informed and performance-based designs. As a result of this practice, most NRC regulations were prescriptive and were applied uniformly to all areas within the regulatory scope. Consideration of the varying risk significance among the areas was limited by the lack of risk-informed methods at that time. The development of new methods has prompted the NRC to initiate a plan for "regulatory improvement" (SECY-94-090).

In a broad sense, risk-informed and performance-based methods can be thought of as a means of providing an alternative option for implementation of regulations that is more efficient in terms of expenditure of resources, while at the same time focusing proper attention on the risk-significant aspects of the regulation. This means may potentially be achieved by an increase in risk-informed discrimination offered by the methodology assessed in this report. The implementation of such a process may be facilitated by the availability of plant-specific PRAs* being performed by utilities in response to NRC Generic Letter 88-20, Supplement 4, on individual plant examinations.

This report presents a technical review and

---

* However, these risk assessments, when used for such purposes, must remain up to date.

analysis assessing the potential for improving the current regulatory system through the use of results and insights gained from risk-informed, performance-based methods. Figure 1.1 is a flow chart depicting the objective and process used in conducting this study. The experience with NRC requirements was reviewed to identify opportunities for the application of risk-informed, performance-based methods, while the availability of these methods was determined in a parallel review. The results of these reviews were used to conduct trial applications of risk-informed, performance-based methods to selected areas of requirements for fire protection contained in Appendix R to 10 CFR Part 50. These applications, or case studies, assess the usefulness of the results and insights from risk-informed, performance-based methods in improving regulatory decisionmaking—in evaluating potential alternative means of implementing NRC fire protection safety objectives—while accounting for uncertainties in the results of these methods.

This report has eight chapters. Chapter 2 describes current NRC regulatory requirements for fire protection in nuclear power plants to establish the foundation for presenting the experience with these requirements. Chapter 3 describes the experience with current NRC fire protection requirements. Alternate methods for fire protection developed since the issuance of Appendix R are presented in Chapter 4. Chapter 5 presents practices and developments outside the NRC and U.S. nuclear industry in nuclear industries abroad, and in other industries in the United States. Chapter 6 presents several trial applications (case studies) evaluating the applicability and usefulness of alternative risk-informed and performance-based methods in improving regulatory decisionmaking accounting for the uncertainties in the results. Potential efficiencies in terms of cost savings that may be gained from applying risk-informed, performance-based methods are presented in Chapter 7. A list of references is given in Chapter 8. Appendices supplement the information in the report.

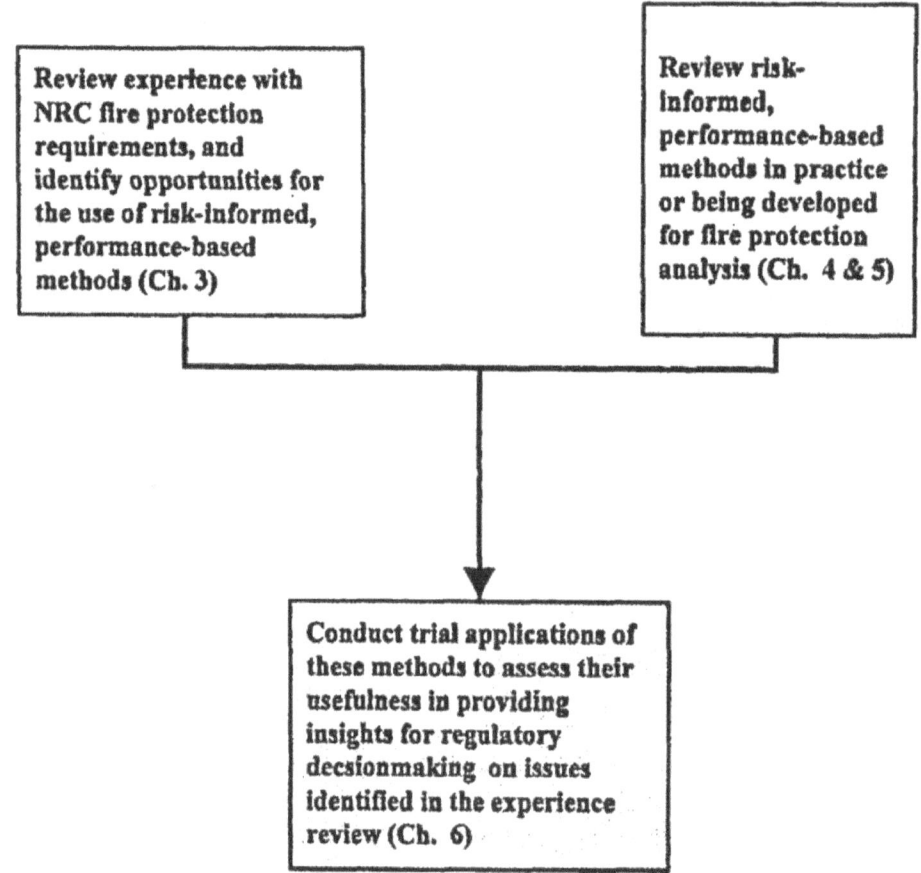

Figure 1.1
Objective of the Study

2 CURRENT NRC REGULATORY REQUIREMENTS

In order to present the experience with NRC requirements (discussed in the next chapter), current NRC requirements are described briefly in this Chapter.

After investigating the 1975 fire at Browns Ferry, the NRC determined that additional specific guidance was necessary to assure that

- the existing fire protection regulations (General Design Criterion 3) were properly implemented

- the established principles of "defense in depth" were applied in defense against fire

Subsequently, in May 1976, the staff issued Branch Technical Position, Auxiliary and Power Conversion Systems Branch, 9.5-1 (BTP APCSB 9.5-1) "Guidelines for Fire Protection for Nuclear Power Plants." The guidance in this document, however, was only applicable to plants that had filed an application for a construction permit after July 1, 1979.

At the time of the Browns Ferry fire, the majority of plants that are operating today were either operating or were well past the design phase and into construction. In an effort to establish an acceptable level of fire protection at these "older" plants, without significantly affecting their design, construction, or operation, the NRC modified the guidelines in the original BTP (BTP APCSB 9.5-1) and, in September 1976, issued Appendix A to BTP 9.5-1, "Guidelines for Fire Protection for Nuclear Power Plants Docketed Prior to July 1, 1976." The NRC then reviewed the analyses submitted by each operating plant against the guidance contained in Appendix A to BTP 9.5-1 and visited plants to examine the relationship between structures, systems, and components important to safety and fire hazards, the potential consequences of fire, and the associated fire protection features.

It is important to note that Appendix R to 10 CFR Part 50 was issued to address only certain "open

issues" raised by the NRC during its review of how operating plants had implemented the guidance contained in Appendix A to the BTP. With the exception of Sections III.G, J, L, and O (which were backfit on all plants regardless of previous approvals granted by the staff), those portions of Appendix A to the BTP that were previously accepted by the staff remained valid. Therefore, Appendix R does not, by itself, define the fire protection program of any plant. For plants licensed before January 1, 1979 (pre-79 plants), the fire protection program is defined by Appendix A to the BTP, the applicable portions of Appendix R (i.e., open issues from Appendix A reviews), and any additional commitments made by the licensee, as stated in conditions of its operating license.

The fire protection programs implemented by the remaining "newer" units were generally reviewed under NRC Standard Review Plan (SRP) Section 9.5-1 (NUREG-0800) and applicable sections of Appendix R (as identified in the plant's operating license).

The operating licenses of pre-79 plants typically contain a condition requiring implementation of modifications committed to by the licensee as a result of reviews conducted under Appendix A to BTP 9.5-1. These license conditions were added by license amendments.

The license conditions for plants licensed after 1979 (post-79 plants) vary widely in scope and content. Some only list open items that must be resolved by a certain date or event (e.g., before startup or before first refueling outage); some reference a commitment to meet sections of Appendix R; and some reference the final safety analysis report (FSAR) or the staff's safety evaluation report, or both.

License conditions did not specify when a licensee may make changes to the approved program without requesting a license amendment. If the fire protection program committed to by the licensee is required by a specific license condition or is not part of the FSAR for the facility, the provisions of 10 CFR 50.59 may not be applied to make changes without prior NRC approval. Thus, licensees may be required to submit license amendment requests even for relatively minor changes to the fire protection program.

To resolve these problems, Generic Letter (GL) 86-10 authorized plants to incorporate the fire protection program and major commitments, including the fire hazard analysis, by reference into the FSAR. In this manner, the fire protection program— including the systems, the administrative and technical controls, the organization, and other plant features associated with fire protection—would be on a consistent status with other plant features described in the FSAR. Also, the provisions of 10 CFR 50.59 would then apply directly for changes the licensee desires to make that would not adversely affect the ability to achieve and maintain safe shutdown. Specifically, GL 86-10 allows licensees to adopt the following standard license condition:

> (Name of licensee) shall implement and maintain in effect all provisions of the approved fire protection program as described in the Final Safety Analysis Report for the facility (or as described in submittals dated _____) and as approved in the SER dated _____ (and Supplements dated _____) subject to the following provision:
>
> The licensee may make changes to the approved fire protection program without prior approval of the Commission only if those changes would not adversely affect the ability to achieve and maintain safe shutdown in the event of fire.

Therefore, plants that have amended their operating licenses in accordance with GL-86-10 may alter specific features of their approved fire protection program provided that

- the change does not otherwise involve a change in a license condition or technical specification or result in an unreviewed safety question (see 10 CFR 50.59), and

- the change does not result in a failure to complete the fire protection program as approved by the Commission.

As with other changes implemented under 10 CFR 50.59, the licensee must

- maintain a current record of all such changes, and

- report all changes to the approved program annually to the NRC Office of Nuclear Reactor Regulation.

Additionally, if the operating license is amended to include this standard license condition,

- The licensee may request an amendment to delete the technical specifications that will now be unnecessary.

- Temporary changes to specific fire protection features which may be necessary to accomplish maintenance or modifications are acceptable provided that interim compensatory measures (e.g., fire watches) are implemented.

Examples of issues that *could* require an exemption (from regulation) or deviation (plant license condition) regardless of license amendment option selected are modifications to

- the level of separation and protection provided for redundant trains of safe-shutdown equipment,

- auto suppression and detection systems, and

- the safe-shutdown methodology approved in the plant's safety evaluation report.

2.1 FIRE SAFETY REQUIREMENTS

The major fire protection requirements for nuclear power plants are the following:

- establishment of a fire protection program

- performance of a fire hazards analysis

- establishment of fire protection features for those areas containing or presenting a fire hazard to structures, systems, or components important to safety

- provision of an alternative or dedicated safe-shutdown capability for areas in which fire protection features cannot ensure safe-shutdown capability

These fire protection requirements have been implemented at all operating nuclear power plants. As described above, the fire protection commitments identified in the plant operating license is a function of vintage and other plant-specific considerations. Pre-79 plants are generally committed to all or portions of Appendix A to BTP APSCSB 9.5-1 and portions of Appendix R to 10 CFR Part 50. Newer plants were reviewed to SRP Section 9.5-1 and the appropriate section of Appendix R.

A brief description of the major sections of Appendix R follows. Section III, "Specific Requirements," is discussed in greater detail to present the necessary background for the case studies in Chapter 7.

I Contains an introduction and discusses the scope of Appendix R.

II Presents the general requirements of Appendix R, including the establishment of a fire protection program, the performance of fire hazards analysis, and the incorporation of fire prevention features into the design and operation of the plant.

III Presents the following specific requirements of Appendix R:

III.A The requirements for fire suppression system water supplies: Two separate water supplies, each consisting of a storage tank, pump, piping, and the appropriate isolation and control valves, are required to furnish the necessary water volume and pressure for the main fire loop.

III.B The requirement for the installation of sectional isolation valves to permit isolation of portions of the main fire loop for maintenance.

III.C The requirement for hydrant isolation valves: These valves permit the isolation of outside hydrants from the fire main for maintenance activities without affecting the protection by the fire suppression system of safety-related or safe-shutdown systems.

III.D The requirement for the installation of sufficient manual standpipe and hose systems so that at least one effective hose stream will be able to reach any location that contains or presents an exposure fire hazard to structures, systems, and components that are important to safety.

III.E The requirement for hydrostatic testing for fire hoses: Hoses stored in outside fire houses must be tested annually; interior standpipe hoses must be tested every 3 years.

III.F The requirement for the installation of automatic fire detection systems in all plant areas that contain or present an exposure fire hazard to safe-shutdown or safety-related systems and components.

III.G The following requirements for protecting the safe-shutdown capability from fire:

(1) Fire protection features must be provided for structures, systems, and components that are important for safe shutdown. One train of systems necessary to achieve and maintain hot shutdown

should be free of fire damage. Systems necessary to achieve and maintain cold shutdown must be repairable within 72 hours.

(2) Separation or protection and, in some cases, detection and automatic suppression are required for redundant trains within the same fire area to ensure that one hot-shutdown train is free of fire damage. Fire areas inside non-inerted containments have additional fire protection options.

(3) Alternative or dedicated shutdown capability is required if the hot shutdown protection/separation requirements of Section III.G.2 are not satisfied, or if fire suppression activities or inadvertent operation or failure of the fire suppression system can damage all redundant hot-shutdown trains. In addition, fire detection and a fixed fire suppression system are required for these areas.

III.H The requirement for the establishment of a fire brigade on site to ensure adequate manual fire fighting capability for all areas of the plant containing structures, systems, and components important to safety. Brigade size, brigade qualifications, and minimum fire fighting equipment are specified.

III.I The requirement for the establishment and maintenance of a fire brigade training program, consisting of periodic classroom instruction, fire fighting practice, and fire drills, to ensure the capability to fight potential fires.

III.J The requirement for the installation of 8-hour, battery-powered, emergency lighting units in all areas in which safe-shutdown equipment must be operated, and for access and egress routes thereto.

III.K The requirement for administrative controls to minimize fire hazards in areas containing structures, systems, and components important to safety. Plant procedures are required to be established, including procedures to control the handling and limit the use or storage of combustibles, govern the use of ignition sources, maintain good housekeeping practices, control plant response to a fire, and define fire fighting strategies to protect safety-related equipment.

III.L The following major requirements address the alternative and dedicated shutdown capability:

(1) The alternative or dedicated shutdown capability must reach cold-shutdown conditions within 72 hours and maintain reactor coolant system process variables within those predicted for a loss of normal ac power.

(2) The functional performance goals of shutdown must be presented.

(3) The shutdown capability must be independent of the specific fire area(s) for which it is being used. It

also must be independent of offsite power for 72 hours.

(4) The systems and equipment necessary for hot shutdown or hot standby must have the capability to maintain such conditions until cold shutdown can be achieved.

(5) The safe-shutdown equipment and systems for each fire area must be isolated from non-safety-associated circuits to ensure that circuit failures will not prevent operation of safe-shutdown equipment.

III.M The requirement that only non-combustible materials may be used for fire barrier cable penetration seals. Qualification testing acceptance criteria are presented.

III.N The requirement that the closure capability of fire doors must be verified periodically to ensure that these doors protect the openings in case of fire.

III.O The requirement that a reactor coolant pump (RCP) oil collection system is provided for non-inerted containments, in order to minimize the likelihood of fires associated with RCP lube oil leaks.

2.2 DOCUMENTATION AND REPORTING REQUIREMENTS

Currently, fire protection requirements are specified in 10 CFR 50.48 and Appendix R to 10 CFR Part 50. Additional guidance is given in NRC generic letters and information notices (stated in Section 2.1). The requirements in Appendix R instruct the nuclear power plant designer on how to provide fire protection that will be acceptable to the NRC, and what actions (e.g., fire brigade training, equipment testing, and inspection procedures) must be carried out to maintain a license to operate.

Each plant maintains a fire protection program plan approved by the NRC staff. If changes are desired in this program plan, approval of such changes is requested from the NRC, and when approved, the changes become a part of the program plan. Fires experienced in the plant that affect safety equipment are reported to the NRC in licensee event reports per § 50.73 of 10 CFR Part 50 and become a part of the permanent NRC record of fire safety experience.

3 EXPERIENCE WITH NRC REQUIREMENTS

This chapter presents a summary of an internal staff review and industry feedback on the experience with NRC requirements discussed in Chapter 2. This is followed by a comprehensive analysis of experience with NRC requirements through a review of exemptions granted to Appendix R. The technical bases for granting the exemptions are identified, and areas in which risk-informed, performance-based methods were or could have been used to provide the basis for the request for approval or granting of the exemptions are presented.

3.1 INTERNAL STAFF REVIEW AND INDUSTRY FEEDBACK

The Regulatory Review Group (RRG), an independent group of NRC staff established by the NRC in 1993, reviewed the fire protection regulations and recommended improvements. The group stated (NRC/RRG, 1993) that "improvements in fire protection material and component performance and the years of fire protection experience and data gained since the issuance of the fire protection rule in 1980, appear to indicate that additional flexibility in the applicable regulations could be allowed without adverse safety impact." The fire protection regulations were also reviewed by the Office of Nuclear Reactor Regulation at the NRC in 1992, and the results of the review were published in the staff report on the reassessment of the NRC fire protection program (SECY-93-143). That report contained a finding that the current requirements and guidelines were developed before the staff or the industry had the benefit of probabilistic risk assessments (PRAs) for fires and before there was a significant body of operating experience. The report concluded that a revised 10 CFR 50.48 (and perhaps the elimination of 10 CFR Part 50 Appendix R) could establish a more reactor-safety-oriented fire protection rule, add appropriate flexibility in some areas, and eliminate the potential for confusion and conflict between 10 CFR 50.48 and Generic Letter 86-10. Additional insights important to fire protection issues in the report are the following :

- Event reports submitted over the last 5 years indicate that typically four or five significant fire events (i.e., those that degrade one or more safety systems or result in a plant transient) will occur each year in all domestic nuclear power plants.

- NRC-sponsored probabilistic fire risk assessments have generally estimated that the core- melt frequency due to fire is currently in the range of 1E-4 to 1E-5 per reactor-year and that implementation of the NRC fire protection requirements has generally reduced the vulnerability due to fire by about 1 order of magnitude. The risk fraction of the total core- damage frequency (CDF) due to fire for the plant can range anywhere from less than 5 percent to more than 50 percent, but for most plants is 20 to 40 percent. Industry studies have indicated that the fire risk fraction of the total CDF is lower. The risk contribution of fires in nuclear power plants is discussed in greater detail in Section 4.4.1.

- Dominant sequences in fire PRA studies typically involve control rooms, control cabinets, emergency switchgear rooms, and cable spreading rooms.

- The Fire Risk Scoping Study (NUREG/CR-5088) concluded that weaknesses in either manual fire fighting effectiveness or control systems interactions could raise the estimated fire-induced CDF by 1 order of magnitude.

- The vast majority of fires are identified and extinguished by plant personnel (including fire watches) and not by automatic detection and suppression systems. The human element is clearly a critical part of the fire safety equation and should be recognized as the first line of defense for mitigating the effects of fire. Fire watches may be more valuable as a mitigating factor than was previously recognized.

- Most fires are of electrical origin, and since electrical fires typically involve significant pre-ignition heating times, they are more likely to be discovered by plant personnel who occupy and tour the different areas of the plant. Also, circuit protective features can interrupt power to faulted circuits and/or the faulted condition can cause control room annunciation before the fire can become fully developed. It is, therefore, important to train plant operators to be sensitive to these scenarios and to respond accordingly.

- Fire event reports indicate that automatic fire suppression and detection systems do not always function properly, and heavy smoke can inhibit manual fire fighting efforts.

- Event reports sometimes describe fire suppression system actuations that cause design deficiencies or maintenance problems to be discovered, such as inadequately sealed components and inadequately sealed fire areas.

- Fire research studies indicate that the 20-foot separation criterion required by Appendix R is not always sufficient in and of itself to protect redundant trains from a single exposure fire. Considerations like this have played an important role in establishing the current defense-in-depth requirements.

- On the basis of inspection experience, it appears that licensees typically maintain their fire protection programs as required by the regulations, and a few licensees actually go beyond the regulatory requirements.

In 1997, the NRC staff completed a special study, AEOD/S97-03 (NRC, 1997) to examine U.S. operating experience through a review of fire events from 1965 through 1994. The report identified the following major findings and conclusions:

- A comparison of fire events in the pre-Appendix R period (1965–1985) with fire events in the subsequent period shows that event frequencies have declined slightly, while the safety significance of events has also been lower. Since the fire at the Brown's Ferry nuclear plant, some fires have been

severe in terms of the magnitude and duration of combustion (such as turbine building fires), but their severity in terms of challenges to safety systems operation has been limited.

- Fire durations during power operations and shutdown conditions were generally short (less than 10 minutes).

- Operating experience indicates that the frequency and duration of shutdown fire events appears to be similar or less significant than for fire events occurring at power operation.

On the basis of two questionnaires in conjunction with formal interviews used to survey industry organizations and the NRC, NUREG/CR-4330 reported on regulations that were suggested for improvement. One of the regulations most frequently cited by the industry was 10 CFR Part 50, Appendix R. The licensees suggested modifying specific parts of Appendix R or guidance for it that contained the following:

- disabling automatic features such as transfer functions and system realignments in order to satisfy the separation requirements for safe-shutdown equipment
- assumptions for transient combustible loads for areas with safe-shutdown components
- the loss-of-offsite-power assumption in the event of a fire
- the use of 3-hour fire barriers regardless of fire loading
- fixed emergency lighting for 8 hours regardless of an assessment of the need for lighting
- no credit for operator action to mitigate the effects of plant fires

The proceedings of the workshop on the program for the elimination of requirements marginal to safety (NUREG/CP-0129) were also reviewed. During the workshop, several regulatory areas, including the Appendix R fire protection requirements, were examined. Potential areas for regulatory reexamination suggested by industry were fire hose testing, fire brigade training, standard repair operations for hot shutdown, emergency lighting, suppression and detection system surveillance and maintenance requirements, use of fire watches, the 20-foot separation criterion, prescriptive use of 1- and 3-hour fire barriers, loss of offsite power, and the capability to attain cold shutdown within 72 hours.

3.2 EXEMPTION REVIEW

Since the implementation of Appendix R to 10 CFR Part 50, the NRC has issued approximately 900 non-scheduler exemptions from the fire protection requirements (Levin and Kanz, 1995).* These exemptions implement alternative approaches which provide a plant-specific level of fire safety that is considered equivalent to the prescriptive requirements of Appendix R.

In general, the licensees requested most of the exemptions from the technical requirements of Section III.G, "Fire Protection of Safe Shutdown Capability"; Section III.J, "Emergency Lighting"; and Section III.L, "Alternative and Dedicated Shutdown Capability." Table 3.1 presents a summary of the number of exemptions granted and the technical areas in which equivalency was demonstrated and approved by the staff. Details of these exemptions are presented later.

Given the state of the art for PRA and the fire sciences when Appendix R was adopted, a highly prescriptive regulation was appropriate. The flexibility of the exemption process allowed non-compliances to be examined in detail and approved if equivalency could be demonstrated. However, a cost is associated with each exemption request, both for the licensees and the NRC. Furthermore, the exemption process is itself a disincentive for many licensees, especially when there are no precedents. Rather than be subject to this unknown, many licensees opt to make the necessary changes to prescriptively comply with the regulation. Only when the cost of compliance becomes prohibitive does the exemption process become attractive. Thus, the exemptions that are requested and the subsets that are approved tend to be substantive issues that

---

* In addition, approximately 450 deviations from BTP 9.5-1, Appendix A to BTP APCSB 9.5-1, and SRP 9.5-1 have been approved.

require modifications and potential forced outages, and have a significant economic impact. However, other areas, such as surveillance, for which a licensee is in full compliance, may also prove to be economically significant over the life of the unit.

The effort to review exemptions commenced with a review of selected Appendix R documents for background, including the *Federal Register* statements of consideration for the rule; the interpretation of fire protection requirements found in Generic Letter 86-10; IE Information Notice 84-09, in which the NRC staff cited lessons learned from the fire protection inspections of safe-shutdown equipment; SECY-83-269; NUREG/CR-4330; and NUREG/CP-0129, which contains a summary of the workshop on regulations that have marginal to safety requirements. The intent of this review was to identify areas in which industry compliance to the rule was stated as a hardship as evidenced by the results of inspections, exemption requests, surveys, and input during the workshop.

SECY-83-269 summarized approved exemptions for the 1982-1983 time period. More than 88 percent of the 234 exemptions addressed Appendix R, Section III.G. These include

- fuse removal for hot shutdown or repair of equipment that is not immediately needed (III.G.1)

- partial barriers or less than 3-hour rated barriers (III.G.2.a)

- intervening combustible materials within the 20-foot separation required by II.G.2.b, if the quantity was judged insignificant.

- no automatic suppression (III.G.2.b, c), again with a low fire loading and high compartment ceilings

Three exemptions from the emergency lighting requirement (Section III.J) were approved to allow portable emergency lights inside the containment and simple repairs to emergency lighting.

Table 3.1 Appendix R Technical Exemptions Granted by the Staff

App. R Section	Technical Area	No. Exemptions	Remarks
III.	Specific Requirements		
III.A	Water Supplies	1	
III.E	Hose Testing	1	
III.F	Automatic Fire Detection	14	
III.G.1	Fire Protection Features	2	
III.G.1.a	One Train of Safe-Shutdown Systems Maintained Free of Fire Damage	11	
III.G.1.b	Systems Necessary To Achieve Cold Shutdown Can Be Repaired Within 72 Hours	4	
III.G.2	Redundant Trains of Systems Necessary To Achieve and Maintain Hot Shutdown Outside of Primary Containment	175	
III.G.2.a	3-Hour Fire Barrier	164	
III.G.2.b	20-Feet of Spatial Separation With Automatic Suppression and Detection	129	
III.G.2.c	1-Hour Fire Barrier With Auto Detection and Suppression	122	
III.G.2.d	Inside Containment—Horizontal Separation of More Than 20 Feet	21	
III.G.2.e	Inside Containment—Auto Detection and Suppression	6	
III.G.2.f	Inside Containment—Radiant Energy Heat Shields	7	
III.G.3	Fire Detection and Suppression for Areas Requiring Alternative or Dedicated Shutdown Capability	139	Most plants requested an exemption from auto-suppression in the main control room
III.H	Fire Brigade	1	
III.J	Emergency Lighting	39	
III.L	Alternative and Dedicated Shutdown Capability	36	
III.M	Penetration Seals	4	
III.O	Reactor Coolant Pump Oil Collection System	24	Majority of exemptions were associated with collection tank capacity
	Total	900	

Several exemptions from Section III.L (alternative shutdown) were granted to allow licensees to achieve cold shutdown in more than 72 hours provided that onsite power was used. Requests for exemption from the loss-of-offsite-power requirement were denied.

Some exemptions from Section III.O for reactor coolant pump oil collection systems were approved because of small quantities of oil or the use of non flammable fluid in the pump coupling.

The FIREDAT computerized database (Levin and Kanz, 1995) contains NRC-approved deviations and exemptions granted to licensees from the criteria contained in NRC guidelines on fire protection, namely, Branch Technical Position (BTP) APCSB 9.5-1, Appendix A; BTP CMEB 9.5-1 (NUREG-0800); and 10 CFR Part 50, Appendix R. FIREDAT is used to update the SECY-83-269 exemption summary to early 1994.

The database has identified 1273 approved exemptions and deviations as follows:

Appendix R	=	900
BTP 9.5-1	=	355
Appendix A	=	96
		1351*

As shown in Table 3.1, Sections III.F, G, J, L, and O account for most of the exemptions granted from Appendix R.

Section III.F, "Automatic Fire Detection," has 14 approved exemptions. These approved exemptions address plant areas containing safety-related equipment that lacks automatic fire detection systems. The majority of these exemptions were approved on the basis of low combustible loading in the area and a qualitative assessment of limited damage if a fire were to occur. Other approved exemptions credit fire detection capability within the area (partial) or in

---

* The discrepancy between the 1273 total and the detailed breakdown of 1351 is attributable to those cases in which one exemption or deviation is applicable to multiple requirements or guidance. i.e., Appendix R, III.J and BTP APCSB 9.5-1.

adjacent zones.

As indicated in the earlier discussion of SECY-83-269, the bulk of the approved exemptions (up to 1983) were for the separation requirements of Section III.G. As shown in Table 3.1, Section III.G continues to account for about 87 percent of all approved exemptions in the FIREDAT database.

Section III.G.1.a requires that the installed fire protection features be capable of limiting fire damage so that one train of systems necessary to achieve and maintain hot-shutdown conditions is free of fire damage.

The FIREDAT database has identified 11 approved exemptions for Section III.G.1.a. The approved hot-shutdown repairs range from simple low-voltage fuse pulling (to prevent spurious operation) to more complex actions that involve lifting leads and attaching jumpers to permit local equipment operation. A sample of the approved exemptions is discussed below (the bases for the approval of the exemptions presented here are derived from the review of the safety evaluation reports):

*   The Dresden licensee received approval for manual recovery actions [8908220394]**, including fuse removals, fuse replacements, tripping circuit breakers, opening disconnect switches and load shedding. The established plant procedures for these actions as well as licensee controls for fuse replacement (i.e., location, accessibility, surveillance, and operator safety) were considered.

*   FitzPatrick requested an exemption to permit fuse pulling, lifting of leads, and cable cutting, all for low-voltage circuits. The staff approval [8305060548] was limited to high-pressure coolant injection (HPCI) and reactor core isolation cooling (RCIC) fuse pulling. Both recovery actions involved the removal of a single fuse located in the relay room.

*   The Hatch licensee received permission

** [NUDOCS accession number]

[8701070595] for operator action to restore residual heat removal (RHR) pump room cooling, RCIC pump and room cooling, and the diesel generator voltage regulator. The RHR and RCIC room cooling manual actions were estimated to take about 20 minutes each. These actions have a time window of about 4 hours before room temperatures reach the design limit. In the case of the voltage regulator for the diesel generator, its function can be restored in 15 minutes by opening links and installing jumpers. The time available to perform this action is ½ hour. In order to perform this task, a dedicated operator will be immediately dispatched to the diesel generator building upon the loss of offsite power. The licensee has also committed to store the tools necessary for the repairs in locked boxes and cabinets.

- Vermont Yankee [8612090830] received an exemption to permit RHR and RCIC fuse replacement to achieve and maintain hot shutdown. The RCIC system is required to be operational within 43 minutes of reactor scram, and the RHR system is required to be operational within 3 hours of reactor scram. In either case, it is unlikely that all fuses would be damaged. However, in either case, all fuses could be replaced in less than 20 minutes, and two sets of spare fuses are readily available at the locations needed.

  The exemption also allows the operators to connect a backup battery charger to the alternate shutdown battery in the event of a fire in the cable vault. The post-fire loads are not expected to discharge the battery before 24 hours, and the alignment of the backup battery charger is considered to be a routine action.

- The Pilgrim licensee received permission to replace five control power fuses and to assume local control of five valves for torus cooling [8901180397]. The staff considered that the 2-hour time period before torus cooling was necessary was much longer than the estimated 20 minutes for fuse replacement, and that the detailed procedures and operator training would ensure success of

the action.

- The use of gasoline-powered fans for charging pump cubicle and emergency switchgear room ventilation was approved for the Beaver Valley licensee [8303290263]. The fans would be set up and operated by the fire brigade. A 1-hour to 2-hour time period is available before high ambient temperatures could damage critical equipment.

- The Big Rock Point fire analysis assumes the loss of instrument air in certain fire areas due to a loss of service water for compressor cooling. This disables two air-operated valves that must be opened to supply makeup water to the emergency condenser. The licensee received approval [9002220554] to manually recover instrument air. This involves cross-connecting the demineralized water system to a portion of the service water system with a cooling water hose. The hose is stored on site. This recovery action appears to be formalized in a procedure and is estimated to take about 10 minutes to complete. The available time to establish emergency condenser makeup is about 4 hours.

- Sequoyah received permission* to use local control of a main control room air-handling unit (AHU) [8606110363]. This involves lifting leads in a 480-V shutdown board, installing a jumper, and replacing the necessary control fuses. The manual actions are proceduralized, and are estimated to require about 1 hour. Adequate personnel are available to perform the required actions within the estimated 5-hour window. The staff also credited the auxiliary building fire protection features, which should reduce any major fire damage to the cabling and components of the control room ventilation system.

In general, the approved exemptions have the following characteristics:

---

* Permission was granted for a deviation from guidance since Sequoyah was not required to comply with Appendix R and did not need to be exempted.

- Relatively simple repairs that typically take 20 minutes or less to complete.

- The necessary tools and material are controlled and readily available. The time available to complete the repair provides reasonable assurance of success.

- The repairs are formalized in the plant procedures.

- The shift staffing has been examined to ensure sufficient personnel are available.

- The repair environment and the nature of the repair do not endanger plant personnel.

Section III.G.2 applies to fire areas that contain redundant trains of systems necessary to achieve and maintain hot-shutdown conditions. This section allows several methods to ensure that cables, equipment, and associated circuits of at least one redundant train of systems necessary to achieve and maintain hot shutdown is free of fire damage. These methods include the use of 3-hour-rated barriers (III.G.2.a), 20-foot separation (III.G.2.b), or 1-hour-rated barriers (III.G.2.c). The latter two alternatives require that fire detectors and an automatic fire suppression system be installed in the area.

Section III.G.2.a permits the use of 3-hour-rated barriers to separate redundant trains. Structural steel forming a part of, or supporting, these barriers must also be protected to provide a fire resistance equivalent to that of the barrier.

The FIREDAT database lists 164 exemptions to III.G.2.a. Exemptions were granted for unrated components (such as water-tight doors or steel hatches), partial 3-hour barriers, barriers with unprotected openings, less-than-3-hour barriers, or components (such as dampers or doors) with less-than-3-hour ratings.

The staff frequently cited low fire loading in its review of the licensee's exemption request. One candidate area for examination using performance-based methods (i.e., fire modeling) are configurations that have complete, albeit less-than-3-hour-rated, barriers with low combustible

loading. An assumption of the amount of transient combustibles is typically embedded in this determination. Several examples are discussed below.

Peach Bottom received several exemptions for less-than-3-hour-rated barriers. In one instance [8503260032] a 1½-hour-rated damper was deemed to provide equivalent protection for a switchgear room. The area has a fire detection system, manual hose stations, and portable fire extinguishers. The fixed combustible loading is approximately 27 minutes using the equivalent fire severity method (NFPA, 1991) which correlates a fire loading to an equivalent fire severity approximately equivalent to that of test under standard ASTM E-119 curve for a specific period*. The fire detectors will reasonably assure that a fire will be discovered in its incipient stage. Although the staff anticipated a time delay between the receipt of the initial fire alarm and the arrival of the fire brigade, the low fire loading provides reasonable assurance that the 1½- hour damper will provide adequate protection in the barrier.

The licensee also received exemptions for several concrete block walls separating emergency switchgear and battery rooms [9110220275]. The walls have a fire-resistance rating of only 2 hours. The maximum combustible loading is 28,800 Btu/ft^2 with an equivalent fire severity of 19 minutes. Automatic smoke detectors are installed in each of the rooms.

In its approval of similar exemption requests for Pilgrim [8810180045], the staff noted that it previously reviewed and approved the concept of fire protection engineering evaluations to document the adequacy of fire protection measures at Pilgrim when the existing configuration was not otherwise in strict compliance with Appendix R.

Other licensees have received similar exemptions.

---

* The method and data for barrier ratings and how they are applied can be found in NUREG-1547. This report also presents findings regarding the acceptability of the equivalent fire severity method.

Table 3.2   Additional Approved Exemptions From Section III.G.2.a

Plant	Reference	Conformance Issue	Fire Severity (Minutes)
Salem	8907270300	1½-hr doors	1 to 46
		1½-hr dampers	1 to 46
		1-hr ventilation ducts	1 to 46
Duane Arnold	8401170530	1½-hr doors	8 to 23
		1½-hr dampers	8 to 23
		2-hr doors	6 to 24
Grand Gulf	9109060092	2-hr walls	15 to 30

Some of these approved exemptions are summarized in Table 3.2 (above).

The staff's key consideration appears to be the preservation of the defense-in-depth concept. The combustible loading in the areas adjacent to the nonconforming barriers is appreciably lower than the installed barrier, generally by a factor of 2 or more.*

In addition to low fire loading, exemptions for partial barrier designs credited fire detection (sometimes with autosuppression), barrier location, or room geometry. For those areas without barriers, low fire loading in conjunction with installed detection, or fire detection with automatic suppression, were generally cited as providing reasonable assurance that at least one redundant hot shutdown train will be free of fire damage.

Although not explicitly cited in the FIREDAT database, it appears that at least two licensees may have used fire modeling in support of their exemption requests for unprotected structural steel forming part of, or supporting, a required fire barrier. For example, the Susquehanna licensee submitted calculations demonstrating that steel in

areas with automatic suppression, or subject to a cable fire from two or fewer trays, could not be raised to its failure point [8908170037].

Section III.G.2.b is a second means to ensure that cables or equipment of redundant trains necessary to achieve and maintain shutdown be free of fire damage. This section requires separation of cables, equipment, and associated circuits of redundant trains by a horizontal distance of more than 20 feet with no intervening combustibles. In addition, fire detectors and an automatic suppression system must be installed in the area.

Most of the exemptions to this section address cases with fewer than 20 feet of separation (or separation with intervening combustibles) or no automatic suppression in the area or both. Low fire loading in the area and a fire detection system were major considerations cited in the exemptions. These factors would allow a fire to be discovered and extinguished before a redundant train was damaged.

Fire modeling was used to support at least two exemption requests from Section III.G.2.b. The FitzPatrick licensee used a fire model to verify adequate separation between redundant trains without taking credit for the installed detection and suppression systems. The exemption was approved on February 1, 1984 [8402230438]. Another exemption from the 20-foot separation criterion was approved in 1991. It involved

---

* Exemptions have been granted for fire severities that approach the rating of the installed barrier on the basis of such additional considerations as installed automatic suppression.

process monitoring instruments in the containment air room that are necessary for safe shutdown, but where the redundant systems are not protected by the 20 feet of separation, barriers, or installed protection systems (sprinklers). The computer program HAZARD I was used to show that the largest credible fire, a self-initiated (electrical overload) fire in one of the redundant cable trays, would not increase the temperature of lower layer air enough to cause damage to the instruments or their process tubing, and that the upper air layer would not descend to the vicinity of the instruments.

Section III.G.2.c provides another compliance method to protect safe-shutdown capability. One of the redundant shutdown trains is enclosed in a 1-hour-rated fire barrier. Fire detection and automatic fire suppression is also required in the area.

The FIREDAT database has 122 approved exemptions. Exemptions were granted for partial or no 1-hour barriers, less than 1-hour barriers, partial or no autosuppression, and combinations of these.

The lack of areawide automatic suppression was the issue in many of these exemptions. The staff generally cited low in situ combustibles using a reasoning that is similar to the approved III.G.2.a exemptions for barriers that are less than 3-hour rated.

Exemptions were also approved for configurations that had barriers that were less-than-1-hour rated. Again, low *in situ* combustibles were a major consideration.

Table 3.3 presents examples of these exemptions.

Several exemptions were approved for areas without 1-hour barriers or area wide automatic suppression. As before, low fire loading, fire detection, and, as appropriate, partial auto-suppression were generally cited. A limited number of exemptions were issued for areas that do not have 1-hour barriers or any automatic suppression. These approaches credit operator action in a process that is conceptually similar to PRA recovery modeling. For example, the Farley licensee was granted exemptions for various fire areas [8701080637] that (1) credit detailed fire procedures and operator action to regain control of the service water system, a pressurizer power-operated relief valve (PORV), charging pump miniflow; (2) establish reactor coolant pump (RCP) seal injection; (3) isolate various sample lines; etc. Another approved exemption of this kind was for Indian Point Unit 2 [8703110139]. The licensee committed to provide portable exhaust fans as an alternative means of cooling pump rooms.

In accordance with Section III.G.3, if the protection requirements of Section III.G.2 cannot be satisfied for the area, room, or zone under consideration, alternative or dedicated shutdown

Table 3.3 Sample Approved Exemptions From Section III.G.2.c

Plant	Reference	Conformance Issue	Installed Features	Fire Severity (Minutes)
ANO-1	8304060505	No autosuppression	1-hr barrier, fire detection	Negligible (in situ)
Salem	8907270300	No autosuppression	1-hr barrier, fire detection	< 10
Rancho Seco	8301140522	No autosuppression, lack of 1-hr barrier*	30-min barrier, fire detection	< 7
Sequoyah	8606110363	Lack of a 1-hr barrier, area wide suppression	40-min barrier, fire detection, partial auto-suppression	Negligible (overall)

* Barrier is calcium silicate, rated for 30 minutes.

capability is required. In addition, this section requires fire detection and fixed fire suppression for the fire area (i.e., the area, room, or zone under consideration).

Most of the approved exemptions addressed the fixed fire suppression requirement for the main control room. The primary considerations in granting these exemptions were low fire loading, partial or full fire detection, and the fact that the control room is continuously manned.

Exemptions from the fixed suppression requirement were also granted for such other plant areas as electrical penetration rooms. Low fire loading and fire detection capability were generally credited in these exemptions. The rationale is that any fire that started would propagate slowly, allowing ample time for detection and manual suppression.

An exemption was identified for the FitzPatrick plant [8305060553], which is conceptually similar to a PRA recovery model. The exemption permits low-voltage fuse pulling, lifting of leads, and cable cutting in the cable tunnel to mitigate the effects of fires in certain ar..as.

H.B. Robinson received an exemption for its service water pumphouse [8312140199]. The area does not comply with Section III.G because it does not have an automatic suppression system, 20 feet of separation or 1-hour barriers, and an automatic detection system. There is no alternate shutdown capability for this area. The licensee justified this alternative on the basis of the following considerations:

- manual fire fighting capability

- television camera surveillance of the area by security personnel in lieu of fire detection

- low combustible loading—An analytical model was employed to show that the magnitude of any exposure fire needed to damage redundant components is significantly higher than reasonably expected.

Other approved exemptions have cited auto-suppression system waterflow alarms, in lieu of installed detection systems.

Section III.J requires emergency lighting units with a minimum 8-hour battery-powered supply for all areas needed for the operation of safe-shutdown equipment and for the access and egress routes thereto. The FIREDAT database has 39 approved exemptions. The lack of emergency lighting in certain plant areas comprised most of the exemptions. Although these areas were typically inside the containment or in the yard, some exemptions applied to indoor areas outside the containment. In reviewing exemption requests from the III.J lighting requirements, the staff considered the timing of the manual actions that require emergency lighting. Many of the actions are for cold shutdown and can be performed several hours after a fire-induced loss of power. For example, ANO-1, received an exemption [8811070033] for a lack of emergency lighting indoors on elevation 317 because the need to access safe-shutdown equipment in that area occurs after the 8-hour battery-powered emergency-lighting time frame expires. South Texas [NUREG-0781, Supp. 4] received a similar exemption for the lack of battery-powered emergency lighting inside the containment based on the need for access in the 8–10 hour time frame.

St. Lucie 2 [8612100269] received permission to use dedicated portable lights for manual operation of the shutdown cooling valves inside the containment. Turkey Point 3 and 4 received a similar exemption. In that document [8404230366], the staff noted that additional personnel will be available during this period to carry and position the lights.

Several licensees received exemptions to use security lighting as an alternative to Section III.J lighting for the yard. The security lighting was generally powered by a dedicated security diesel generator (ANO [8811070033], Haddam Neck [8712210060]). The Hatch licensee's [8701070595] security lighting will not be available if offsite power is lost. Hatch has dedicated engine-driven portable lights to illuminate the required areas in the yard as a backup.

There have been several exemptions issued for the use of portable lighting, both indoors and in the yard. Quad Cities [9106060039] can use portable lighting to read the suppression pool level sight glasses. This action is expected to be required in no sooner than 3 hours. H.B. Robinson received permission to use portable lighting in several areas of the plant, both inside and outside the containment [8807120348, 8708060038, 9210160190]. As discussed above, St. Lucie and Turkey Point have received permission to use portable lighting for valve alignments inside the containment.

Portable lighting in the yard is primarily used for operator access and egress (Davis-Besse [9004240205], Millstone 1 [8708060275]). In addition, Brunswick [8701020203] uses portable lighting to read gauges in the yard. Haddam Neck uses portable lighting to supplement security lighting for access/egress and manual valve operation.

Several licensees have received permission to use hardwired lighting systems instead of battery-powered emergency lighting. With the possible exception of Fort St. Vrain [8805240108], these exemptions address specific fire areas. Davis-Besse received an exemption [9004240205] for the use of its essential lighting system in parts of the auxiliary and turbine building for fires in the control room and the cable spreading room. A fire in any area outside the control room will not cause the loss of both divisions of emergency lighting. Diablo Canyon [NUREG-0675, Supp. 23] has also received credit for hardwired lighting systems. As discussed below, Hatch uses its hardwired emergency lighting systems as a backup to its 2-hour-rated battery-powered lights in the control room.

Although there are some small inconsistencies, primarily with regard to the use of portable lights and manual actions, the staff has approved exemptions to the emergency lighting require-ments if the alternative could provide enough illumination to facilitate the task and was reliable. In general, the adequacy of the illumination level was verified in the field. Routes of travel were examined for obstructions and tripping hazards. The evaluation of reliability was dependent on the alternative. The design and routing of hardwired systems were reviewed to ensure availability for each fire area that was credited. Portable lights require a program to ensure both the availability and operability of the flashlights, when needed.

Two exemptions [8507160041, 8701070595] were for control room lighting for 90- or 120-minute batteries for the Duane Arnold and Hatch plants, respectively. For example, at Hatch, the emergency lights in the control room are designed to be powered initially from the station batteries and later transferred to the emergency diesel generators after they are started. The emergency lights are designed so that a fire in any area outside of the control room or cable spreading room would not result in the loss of both divisions of emergency lighting. The feeder circuits outside of the control room and the cable spreading room have divisional separation equivalent with the separation requirements of Section III.G.2 of Appendix R. Therefore, the emergency lighting would be supplied with diesel-driven ac power prior to battery depletion.

The Beaver Valley licensee [8701070595] received an exemption which allows the use of 2-hour-rated emergency lighting in the fire brigade room. This room is used as a staging area for alternate shutdown procedures and is expected to be used less than 30 minutes.

Section III.L of Appendix R states the requirements for alternative or dedicated shutdown capability. This capability is required when the separation requirements of Section III.G cannot be satisfied. The majority of the exemptions from Section III.L were granted in the three areas discussed below.

Several exemptions were granted from the requirement to maintain reactor coolant system process variables within those predicted for a loss of normal ac power (LOSP). These exemptions were for boiling-water reactor (BWR) licensees that generally employed rapid reactor pressure vessel depressurization as part of their alternative shutdown capability. This rapid depressurization can temporarily lower the vessel level below the core. The basis of these approvals was the assessment that the fuel rod cladding would remain intact, despite a temporarily depressed water level. PRAs also typically use fuel cladding integrity as the measure of successful core cooling in those cases in which marginal mitigation capability is available.

The unavailability of a complete set of process variable readings for the alterative shutdown process also accounted for several approved exemption requests. The considerations cited in these approvals were the availability of reading material that could provide similar information or an assessment that the subject parameters were not necessary to assure a safe and stable shutdown condition.

The last major subject of approved exemptions from Section III.L concerned the capability to reach cold shutdown (with onsite power only) within 72 hours. Six PWR licensees received exemptions from the 72-hour requirements. As part of the approval process, the NRC qualitatively assessed the safety significance of using nonstandard system alignments over a protracted time period to reach cold shutdown.

Section III.O of Appendix R requires a reactor coolant pump (RCP) oil collection system for non-inerted containments. The majority of the 24 approved exemptions from this section were for systems with collection tanks that were not sized for the entire inventory of all the RCPs. In general, those systems could contain the lubricating oil contents of one pump. The exemptions were granted for RCP lubricating oil systems that are seismically qualified and, therefore, subject to small random leaks.

3.3 CONCLUSION

The following conclusions are drawn on the basis of the review of exemptions to Appendix R granted by the staff:

- The justifications submitted by licensees for the request for exemptions, and the technical bases used by the staff for granting the exemptions, were primarily qualitative analyses of combustible loading and effect

based on engineering judgment; in a few cases, licensees submitted quantitative analyses using fire models as part of the justifications for the exemptions.

- Qualitative analyses and arguments similar to those in recovery models in PRA human reliability analysis (HRA) were used in several submittals for exemptions; however, quantitative PRA or HRA analyses were not submitted at that time.

- Most of the exemptions are in technical areas amenable to the use of risk-informed, performance-based methods that have been developed since the issuance of Appendix R and exemptions granted to that regulation, e.g., fire PRA including HRA analysis, and modeling the dynamics of fire effects.

The opportunities for the use of risk-informed, performance-based methods are discussed further in Chapter 6. Trial applications are presented to evaluate the usefulness of the results and insights from these methods in improving regulatory decisionmaking, on issues that were the subject of past exemptions presented above.

On the basis of the review of the experience and exemptions, the following issues were chosen for further analysis and for trials of risk-informed, performance-based applications:

- emergency lighting (Section III.J), specifically the requirement for an 8-hour lighting duration

- the 72-hour cold-shutdown capability and requirement of Sections III.G.1 and III.L.5

- surveillance requirements for fire detectors

- surveillance test duration for emergency lighting

- the 20-foot safe separation requirement of Section III.G

- the loss-of-offsite-power requirement for alternative or dedicated safe-shutdown capability (Section III.L)

4 ALTERNATE METHODS DEVELOPED SINCE ISSUANCE OF APPENDIX R

This chapter summarizes fire probabilistic risk assessment (PRA) and modeling methods that have been developed and used by the NRC and the U.S. nuclear industry for conducting PRA studies, and by licensees for conducting individual plant examinations for external events (IPEEES) in response to NRC Generic Letter 88-20, Supplement 4. The results of PRAs and the IPEEEs are currently not used to support regulatory decisionmaking for the implementation of NRC fire protection regulation, but have been limited thus far to determine if specific vulnerabilities to fires exist in plants. The methods that have been developed and used are described below, followed by a discussion later in the chapter of the current uncertainties associated with the methods. This chapter also describes the methods and some findings from their use in the past, and summarizes the experience from their use as related by the users. These methods were also applied specifically for the purposes of this study, i.e., to assess their usefulness in improving regulatory decisionmaking in evaluating alternative methods for implementing current fire protection requirements. A critical analysis of the usefulness of the results and insights from the applications for this study, in light of the uncertainties associated with the methods, are presented in Chapter 6.

4.1 FIRE PROBABILISTIC RISK ASSESS-MENT METHODOLOGY

Internal fire PRAs typically follow a two-phase approach. In phase 1, a screening analysis is performed to identify the important fire locations and screen out those areas that are not risk significant. In phase 2, a detailed analysis is performed for the important fire scenarios. The results of a fire PRA are usually obtained from the logic trees and models developed for internal event PRAs. The input probabilities to the PRA models are determined from a performance evaluation of the fire scenarios (propagation, damage, and suppression) and an analysis of fire frequencies. The performance evaluation models

used in fire PRAs are usually based on reliability and/or state-transition models for suppression, and on deterministic phenomenological models (e.g., COMPBRN) for fire growth. Summaries of the approach of typical internal fire PRAs (NUREG/CR-2300, NUREG/CR-2258, Indian Point 2 PRA (Consolidated Edison, 1992), Limerick PRA (NUS, 1983)) follow. A fire PRA utilizes the models developed for an internal event PRA.

4.1.1 Identification of Fire Areas, Fire Zones, and Critical Fire Locations

The fire areas and fire zones as defined in the plant's submittal in accordance with Appendix R to 10 CFR Part 50 are used in the screening analysis. First, fire areas and zones (1) that do not contain safe-shutdown equipment and (2) in which a fire will not adversely impact safe-shutdown equipment in other fire areas and zones are eliminated from consideration. In the Appendix R analysis, the safe-shutdown equipment is tabulated for each fire area and zone. This information is used in the screening analysis of the fire areas and zones. That is, all of the equipment in the fire area and zone is assumed to be disabled by the fire. As a result, an accident-initiating event may occur (i.e., a fire-induced transient or a loss-of-coolant accident). The relevant event tree of the internal event PRA is used to calculate the contribution of the fire zone to core-damage frequency (CDF) by using the fire frequency for the area or zone as the initiating event frequency and assuming that all components within the fire area or zone fail. Some numerical screening criterion, such as 1E-08 per year, is used to screen out unimportant fire areas and zones.

Critical fire locations are those locations for which fire accident scenarios would be developed. Their identification inside a fire area or zone requires cable routing information obtained by tracing the cable routing drawings and by performing a walkdown of the plant. The determination of the critical fire locations is based on the effect of the postulated fire at the location and is done subjectively. The criterion is that a postulated fire at the location must cause an initiating event and failure of multiple equipment needed to mitigate the accident.

4.1.2 Estimation of Fire Frequency for Fire Areas, Fire Zones, and Critical Fire Locations

The data collected from the total population of U.S. nuclear power plants are used. A recent compilation of fire incidents at nuclear power plants has been prepared by Houghton in NRC document AEOD/S97-03 (NRC, 1997). The two ways of estimating the fire frequency for a critical fire location are the following:

Area based. The fire incidents are grouped on the basis of the fire area or building in which they took place (e.g., the switchgear room or the auxiliary building) to estimate the fire frequency of the respective fire areas. This overall area-based frequency is apportioned among the fire zones and critical fire locations on the basis of the components within those zones. This approach is sometimes used to estimate the transient fire frequencies.

Component based. The fire incidents are grouped on the basis of the type of component that was involved in the initiation of the fire (e.g., cables, motor control centers, panels, and pumps) and are used to estimate the fire frequency by component type. The frequency with which a fire occurs at a critical location is determined by prorating the plantwide frequencies of each component type within a critical location.

Data analyses and estimation techniques play an important role in performing this type of evaluation. Several personal-computer-based tools are available for these types of analyses (e.g., Azarm and Chu, 1991).

4.1.3 Fire Damage and Suppression

COMPBRN IIIe* and the worksheets in the FIVE methodology (discussed later) have been widely used for predicting the fire propagation times. Both divide the compartment into at least two zones (an upper layer of hot gas and a lower layer). The fire and its plume may be separate zones or may be included in the upper zone. The gas layers are assumed to be well mixed. Other zone models (e.g., CFAST) that are being used to support non-nuclear plant applications (e.g., fire regulation of buildings) exist and, depending on the application, have different strengths and weaknesses. These models are discussed in the next chapter.

COMPBRN III is a deterministic fire hazard computer code designed to be used in a probabilistic analysis of fire growth in a compartment. Its primary application to date has been in the assessment of fire risk in the nuclear power industry. COMPBRN III follows a quasi-static approach to simulate the process of fire growth and the resulting thermal hazard (including temperature and hear fluxes) during the pre-flashover period in an enclosure. The dimensions of the compartment, location, quantity of fuel, layout of cables, locations and sizes of doorways, and ventilation rates through ventilation ports are user specified.

Possible outputs of COMPBRN include the total heat release rate of the fire, the average temperature and thickness of the hot gas layer, the mass burning rate for individual fuel elements (affected by thermal radiation from the ceiling layer), and the surface temperature of non-burning elements. The time until the target (e.g., cable tray) reaches its damage temperature is the time available for fire suppression. Fire suppression data can be used to determine a probability distribution for the time to suppression, and the probability that a fire is not suppressed before it propagates can be determined using such a curve. Siu and Apostolakis (1986) give more detail on how fire detection and suppression can be modeled in a fire PRA.

---

*COMPBRN IIIe is an improved version of COMPBRN III (EPRI, 1991)

Experimental data from the UL/SNL series (NUREG/CR-3192) are used in Electric Power Research Institute report EPRI NP-7282 to demonstrate the reasonableness of COMPBRN predictions for a representative scenario. The data include doorway flows (driven by the buoyancy of the hot gas), the gas temperature in the hot layer, and cable temperatures.

A number of "field" models for application to fire problems are currently under development. The field model is a complex fluid mechanics model of turbulent flow derived from classical fluid dynamics theory. This type of model solves the fundamental equations of mass, momentum, and energy. In order to facilitate the solution of the equations, the space being analyzed is divided into a three-dimensional grid of small cells. Field models typically use hundreds of thousands of cells or zones; zone models use two or three. The field model calculates the physical conditions (temperature, gas velocity, species concentration) in each cell, as a function of time. The size of the space can range from an area within a room to a large portion of the outdoors (Stroup, 1995). Field models are being used to analyze a number of fire protection issues such as the placement of heat and smoke detectors, and the interaction of sprinklers, vents, and draft curtains. A brief description of CFD codes available or being developed is presented in the next chapter. These codes have not as yet been used in the U.S. nuclear industry.

4.1.4 Fire Event Trees

For a fire at a given critical fire location with the fire either suppressed or propagated, the equipment that will be damaged by the fire is determined. Therefore, the effect of the fire on the plant's capability to mitigate it is defined. An applicable internal event tree can then be modified to model scenario progression. The quantification of the event tree accounts for the frequency of the fire at the location, the probability of fire propagation before suppression, and the availability of alternate equipment for safe shutdown. The fire event tree analysis is similar to that of internal event analysis, except that the impacts of the fire on equipment and operator actions are addressed.

Several tools developed by both the NRC and industry are available for the purpose of such an analysis. The NRC code known as IRRAS (NUREG/CR-5813) is widely used by NRC contractors. Other widely used proprietary computer tools for this purpose are RISKMAN (Pickard, Lowe and Garrick, Inc.), NUPRA (NUS Corporation) and SAICUT (Science Applications International Corporation).

4.2 THE "FIVE" METHODOLOGY

The fire-induced vulnerability evaluation (FIVE) methodology (EPRI TR-100370) is oriented toward uncovering plant fire vulnerabilities. It provides a combination of deterministic and probabilistic techniques, similar to PRAs, for examining a power plant's fire propagation and protection characteristics. The FIVE methodology was developed in response to NRC Generic Letter 88-20, Supplement 4. A utility may choose to conduct a fire PRA or use the FIVE screening methods to conduct IPEEEs in its response to Generic Letter 88-20, Supplement 4. The results of fire modeling worksheet used in the FIVE methodology have been compared (EPRI TR-100443) with data from two series of large scale tests: the FM/SNL series (NUREG/CR-4681, NUREG/CR-5384) and the UL/SNL series (NUREG/CR-3192).

The FIVE methodology is very similar to fire PRA methodology with the following exceptions:

- FIVE uses the progressive screening approach at various stages of evaluation and usually gives full credit (i.e., failures are not evaluated) to the areas that are in compliance with Appendix R, unless additional analyses are deemed necessary by the analytical team (see discussion on p. 6-2 of EPRI TR-100370 on the requirement in Section III.G.2b in Appendix R).

- FIVE provides guidelines to assess the potential for fire propagation across compartments due to failure of barriers and penetration seals. The fire PRA process could also address this issue, but that is generally not the practice.

- FIVE recommends that the self-ignited fire frequency for cables rated according to Institute of Electrical and Electronics Engineers (IEEE) 383 standard be set to zero (in contrast to past PRAs) regardless of the voltage and rated power.

- FIVE provides tables, worksheets, and various equations for fire propagation analyses whereas COMPBRN IIIe is a computer code.

- FIVE provides tables for estimating the availability of automatic suppression and detection systems (p. 10.3-7 of EPRI TR-100370, Table 2). The unavailabilities reported in this reference are more optimistic than those values used in PRAs (Gallucci and Hockenbury, 1981).

- FIVE credits only those systems for which cable routing and evaluation have been performed in accordance with Appendix R. Other systems that may be unaffected by a fire may not be credited if they were beyond the current scope of Appendix R documentation in the plant. In a fire PRA, an analyst may or may not choose to credit those systems.

In 1995, EPRI issued a Fire PRA Implementation Guide (EPRI TR-105928) for use by licensees in conducting the IPEEEs. This guide uses many of the methods and assumptions included in the "FIVE" method. Some assumptions in the PRA Implementation Guide that go beyond those in FIVE have been questioned by the staff.

4.3 PRELIMINARY IPEEE RESULTS

A report (NRC memorandum, 1998) has been developed by the NRC documenting preliminary insights on the results generated and methods used in an initial set of IPEEEs. The initial set of IPEEEs used either the FIVE method, a fire PRA, or a combination of the two methods. The EPRI Fire PRA Implementation Guide was not available to licensees that submitted these IPEEEs before the guide became available as an additional option.

The objectives of the IPEEEs were to identify vulnerabilities to fire events using the methods described above, and to implement cost-effective safety improvements to either eliminate or reduce the impact of these fire vulnerabilities. On the basis of the reviews of an initial set of IPEEEs, the staff has made a preliminary conclusion that most of the licensees whose studies were reviewed have met the objective of the IPEEEs using the methods described above. The report also provides summaries of results, findings and plant improvements reported in the IPEEEs, and additional perspectives related to fire events, and the strengths and weaknesses of the methods used toward accomplishing the IPEEE objectives.

4.4 RESULTS FROM FIRE PRAS

4.4.1 Review of 12 Fire PRAs

Fire PRAs for 12 operating nuclear power plants were extensively reviewed to determine the contribution of fire to annual CDF and to identify dominant fire sequences and important plant areas from a fire perspective. This section summarizes the review; a detailed discussion of each plant is in Appendix B.

As shown in Table 4.1, the CDF as a result of fire-initiated events varies from 2.3E-4 (Big Rock Point PRA) to 8.1E-8 (McGuire individual plant examination for external events (IPEEE)). The fire-initiated CDFs reported in the IPEEEs are generally 1 or 2 orders of magnitude smaller than those reported in earlier PRAs. PRAs typically identify fires in the switchgear room, auxiliary building, control room, and cable spreading room as the major contributors to fire-induced CDF.

The reasons for the differing contributions of fire to the overall CDF were further investigated. The findings are based on a review of four fire PRAs. The four plants selected for this review cover the varying ranges of the fire-initiated CDF in pressurized water reactors (PWRs) and boiling water reactors (BWRs).

(1) The fire PRA issued in March 1981 for Big Rock Point (Consumers Power Company, 1981) (a BWR plant) reported a fire CDF of 2.3E-4 per reactor-year. The large

Table 4.1 Plant Core-Damage Frequency (CDF)

Plant	Total CDF (per RY)	Fire CDF (per RY)	Contribution of Fire to Total CDF	Reference
Indian Point 2*	9.6E-5	6.5E-5	68%	Indian Point 2 IPE (Consolidated Edison, 1992)
Limerick 1	4.4E-5	2.3E-5	53%	Limerick PRA (NUS, 1983)
LaSalle 2	1.0E-4	3.2E-5	32%	NUREG/CR-4832, Vol. 1
Big Rock Point (BRP)	9.75E-4	2.3E-4	24%	BRP PRA (Consumers Power Company, 1981)
Peach Bottom	1.1E-4**	2.0E-5	18%	NUREG-1150, Vol. 1; NUREG/CR-4550, Vol. 4, Rev. 1, Part 3
Seabrook	2.3E-4	1.75E-5	9%	Seabrook PRA (Garrick et al., 1983)
Zion	4.9E-5	4.6E-6	9%	Zion PRA (Commonwealth Edison Co., 1981)
Surry	1.5E-4	1.1E-5	6%	NUREG-1150, Vol. 1; NUREG/CR-4550, Vol. 3, Rev. 1, Part 3
Oconee	2.5E-4	1.0E-5	4%	Oconee PRA (Nuclear Safety Analysis Center, 1984)
South Texas Project (STP)	4.4E-5	4.9E-7	1%	STP IPEEE (Cross et al., 1992)
Catawba 1 and 2	7.8E-5	3.4E-7	< 1%	Catawba IPEEE (Duke, 1992)
McGuire	7.4E-5	8.1E-8	< 1%	McGuire IPEEE (Duke, 1991)

* The Indian Point Unit 2 (IP2) IPE does not contain external events analyses. The fire contribution was taken from a report prepared by EG&G (EGG-2660) in 1991. The data in that report for IP2 were based on a report prepared in the 1980s, and the total CDF was calculated as the CDF from the IP2 IPE (3.13E-5) plus the fire contribution (6.5E-5). The percentage was calculated by this study using these values.
**Total CDF based on seismic analysis using LLNL hazard curves.

contribution of fire to CDF was a result of fires in the cable penetration area inside the containment and the station power room. In both areas, the cables from redundant safe-shutdown trains were routed through the same fire area and adjacent to each other with little or no separation distance (see page VI-25, Consumers Power Company, 1981); therefore, a single fire could have damaged all the cables.

(2) The fire PRA for Limerick (NUS Corporation, 1983), a typical BWR with respect to fire CDF, was issued in April 1983 and estimated a CDF of about 2.3E-5. Self-ignited cable-raceway fires, including IEEE-rated cable fires, account for more than 50 percent of this contribution. Self-ignited fires for IEEE-rated cable are excluded from PRAs being performed as part of IPEEE/FIVE methodologies. The Limerick PRA was conducted for licensing the plant and did not reveal any vulnerabilities that had to be addressed in the licensing process.

(3) For the PWR plants, the fire CDF, shown in Table 4.1, ranges from 1.0E-5 per reactor-year to 1.0E-7 per reactor-year. The Surry plant ((NUREG-1150, Vol. 1) was selected as a representative plant for the higher range (1.0E-5). About 85 percent of the fire CDF in the Surry nuclear power plant is due to fires that result in reactor coolant pump

(RCP) seal loss-of-coolant accidents (LOCAs).

(4) The South Texas Project IPE (Cross et al., 1992) was reviewed as representative of those PWRs with a low CDF (5.0E-7). Unlike the Surry plant, the South Texas Project has positive displacement pumps capable of providing RCP seal injection. Therefore, South Texas Project is not as susceptible to fire-induced seal LOCAs. Excluding RCP seal LOCA, the fire CDFs for Surry and South Texas are comparable.

4.4.2 Comparison of NRC and EPRI PRA Studies

After the development of the FIVE methodology, the Electric Power Research Institute (EPRI) initiated a fire risk assessment (FRA) program to better understand risk due to fire. To meet this objective, the EPRI program provided to its members a set of user-friendly tools (including FIVE), the needed databases, and an approach for performing PRA. To understand the impact of these tools, two existing fire PRAs (Seabrook and Peach Bottom) were requantified (Parkinson et al., 1993). EPRI's approach, in almost all cases, resulted in significantly lower estimates for fire CDF. As an example, the requantification of the cable spreading room at Seabrook resulted in approximately a factor of 400 below that of the Fire Risk Scoping Study (NUREG/CR-5088). A reduction factor of 10 was obtained for ignition frequency; another reduction factor of 15 was obtained for the overall suppression failure probability; and a reduction factor of 2 was obtained for the probability of a fire occurring in a critical area. These reduction factors stem from the following four major differences:

(1) the impact of the EPRI database on ignition frequency (Attachment 10.3 in EPRI TR-100370),

(2) initial fire heat release rate and rejection of the possibility of large transient fire in the cable spreading room supported by the EPRI database

(3) incorporation of modeling uncertainty in damage time calculated by COMPBRN

(4) modeling of various means of detection and suppression

Parkinson et al. (1993) provides a systematic approach for fire risk assessment using FIVE, COMPBRN, and the existing databases. However, large numbers of assumptions, extrapolation of test data, and interpretation of the past fire events are embedded in the approach. Currently, there is no agreement between NRC and EPRI about the validity of these assumptions.

4.5 UNCERTAINTIES

This section presents a description of the common uncertainties associated with fire models and PRAs that have been raised in the past. Key assumptions, methods, or data that are currently said to be the major sources of uncertainty are presented. A critical analysis of trial applications to assess the usefulness of results and insights gained from fire PRA and modeling methods for improving regulatory decisionmaking in light of these uncertainties is presented in Chapter 6. An extensive description of these uncertainties is provided at this point in the report because these uncertainties are cited most frequently as the basis for the very limited usefulness of risk-informed, performance-based methods for fire protection. These sources of uncertainty is critically analyzed in Chapter 6 in terms of its effect on the usefulness of the results and insights gained from the trial applications.

4.5.1 Fire Models

The uncertainties associated with a fire model in a PRA process may be categorized as follows:

(1) uncertainties in the input variables to the fire model and in the parameters used in the model

(2) accuracy of the fire model, excluding any input variability discussed above

The uncertainty distribution, associated with input variables and model parameters (issue 1), is estimated using measurements or monitored data through application of the Bayes method (Kaplan, 1983). Computer software is widely used for these types of uncertainty analyses for both risk-informed and performance-based models. This technology has been utilized for more than a decade in various probabilistic risk assessments and reliability studies. The uncertainties in input variables and the model parameters are propagated through an integrated model using Monte Carlo sampling techniques. Variance reduction techniques and stratified sampling strategies have been extensively used to propagate the uncertainties in an efficient manner. These techniques have already been developed and software developed (e.g., the IRRAS computer code (NUREG/CR-5813) and COMPBRN (EPRI NP-7282)) for fire PRAs.

The accuracy of model prediction, excluding the variabilities of the input and model parameters, is entrenched in code validation. In most cases, simplifying conservative assumptions have been incorporated to reduce the code's development effort and to facilitate the large number of runs required for conducting fire PRAs. Two methods of validation are usually proposed. The first is the comparison of the code predictions to those of another validated code that is more comprehensive and suffers from fewer simplifying assumptions. The other method requires comparison of the code predictions to available measurements obtained through a well-instrumented experiment.

In any case, exhaustive comparisons of the existing codes to either experiments or to a more comprehensive code are not generally feasible because of the large number of case runs that may be necessary or the cost associated with new experiments and/or additional computer runs. Various statistical methods are available to provide an estimate of the inaccuracies of the code prediction using a small set of validation runs. Currently, expert judgments are used in most cases to determine the accuracy of the code predictions in light of the limited experimental data available. One method used in the building industry, albeit informal, aggregates the results of those fire experiments (or actual fire events) that are judged to be representative of the case under

study, in order to refine the code estimates. The aggregation process is based on the weighted mixture of all results. The closer the fire experiment represents the case run, the higher would be its weight. This is also the case for the computer codes for evaluating fire propagation times.

A formal treatment to determine fire model uncertainties is proposed in Appendix C. Several sources of data uncertainties, i.e., parameter uncertainty and uncertainty of initial and boundary conditions are identified. The current treatment of data uncertainties is summarized and different sources of modeling uncertainties resulting from assumptions, approximations, simplifications, and numerical algorithms are discussed. An approach is proposed on the basis of decomposition of uncertainties to the most basic level of modeling and aggregation of the uncertainties using the current uncertainty propagation techniques. A process for decisionmaking under both modeling and data uncertainty is also presented. This proposed treatment could form the basis of research to further define fire modeling uncertainties.

Methods or effects that are currently stated to be the major sources of uncertainty in fire models based on experience and engineering judgment are discussed below. A more detailed review of the features, limitations and uncertainties in fire models can be found in Mowrer and Stroup, 1998.

4.5.1.1 Source Heat-Release Rates

The largest source of uncertainty in fire models is associated with the heat-release rate (Mowrer and Stroup, 1998). The phenomenological modeling of the combustion process and heat release is extremely complex and in an early research stage. Experimental data are widely used and provided as input to fire models, and large uncertainties are associated with this input because of the inability to accurately correlate experimental data to the fire source of concern. The heat-release rate is the driving force for the plume mass flow rate, the ceiling jet temperature, and finally, the hot layer temperature that is driven by energy balance. The fire heat-release rate is dependent on the initial fire size, the growth of fire by propagation and ignition of additional combustibles, and the heat-release rate from these additional combustibles.

4.5.1.2 Multi-Compartment Effects

It has been stated that a source of uncertainty for certain applications is related to the number of compartments analyzed by the model, and the compartment geometry used in the experimental validation. The COMPBRN code is a single-room model that assumes a small (pre-flashover) fire in a large compartment. Currently, fire probabilistic risk assessments (PRAs) do not consider the fire propagation across fire-rated structural barriers and seals. Generally, PRAs assume that the probabilities of such events are negligible, considering the large size of, and the slow burning materials (cables) in, the compartments of nuclear power plants. PRA analysts sometimes consider that smoke propagates across compartments as a result of damper failures, especially if smoke-sensitive equipment is in the adjacent room.

4.5.1.3 Effects of Ventilation

In certain applications, the effects of mechanical ventilation may be important. Most fire models have difficulty in accurately predicting the effects of mechanical ventilation on fire development and the corresponding effects on the fire compartment(s) and contents. COMPBRN has this feature; however, the experimental validation is lacking. In contrast to COMPBRN, where vent flow is calculated using empirical equations, CFAST (discussed in Chapter 5) utilizes Bernoulli's solution for the velocity equation. This solution is augmented for restricted openings by an empirically based flow coefficient. Forced ventilation is treated as constant flow rate in COMPBRN, whereas in CFAST, the forced-ventilation mass flow rate varies with square root of pressure drop. Nuclear power plants in the U.S. are typically multi-room windowless structures of various sizes and are provided, exclusively, with forced-ventilation systems. Neither COMPBRN nor CFAST is experimentally validated for such configurations.

4.5.1.4 Structural Cooling Effect

Considerable cooling effect can come from the masses of cable trays, ventilation ducts, and

piping in the upper part of compartments in nuclear power plants. Zone models have not been used to calculating the heat transfer by convection from the gas in the hot layer to these structures. Therefore, the gas in the hot layer could actually be much cooler than calculated.

4.5.2 Parameters Important for Calculating Fire Risk

4.5.2.1 Fire Ignition Frequency

Large uncertainties are reported in the estimated fire frequencies for the control room, cable spreading room, and switchgear rooms, primarily because the data are quite sparse. Four fires have occurred in the control room, but all were small with mean duration of 2.5 min (NRC, 1997) and could be extinguished without any need for evacuation. On the basis of these data, the control room fire frequency has been estimated to have a 90-percent confidence range of 1.0E-6 to 7.0E-3 per reactor-year. Similar uncertainty ranges have been reported for the cable spreading room and the switchgear room (NUREG/CR-4550, Vol. 3, Rev. 1, Part 3 and Vol. 4, Rev. 1, Part 3).

The impact of underreported and event screening on the uncertainties of fire frequencies used in fire PRAs has also been raised as a concern. A detailed discussion of this concern and the uncertainties associated with fire ignition frequencies as a result of underreporting and event screening is presented in a recent review sponsored by the NRC (Azarm, 1998). This study concluded that small fires that cause little or no property damage or component failure may not be completely captured by generic databases. The potential impact that small fires could have on risk insights from fire PRAs was investigated. It was concluded that the level of detail in PRA models dictates what fire events should be considered for estimating the initiator-event frequency. More-detailed PRA models reduce variability in the estimated risk, but require more extensive data on fire occurrences. For current state-of-the-art PRAs and the associated level of detail, the available generic databases should be sufficient for obtaining generic risk insights as opposed to detailed plant-specific results.

4.5.2.2 Reliability and Effectiveness of Fire Detection and Suppression

Automatic detection and suppression systems have been backfitted in nuclear power plants and, in some cases, automatic fire detectors and suppression heads may be obstructed by such structures as cable trays, piping, and ducts. How obstructions quantitantively affect the effectiveness of these automatic features is currently unknown.

The response time of fire detectors may be affected by the presence of obstructions. The slower the detector's response time, the larger the size of the fire by the time of detection; so early detection can be important. It might be important to assess the capability of current codes in estimating the detector response.

The zone models calculate the depth and temperature of the ceiling layer as a function of time, but they ignore the transit time for the gas from the fire to rise and mix with the ceiling layers. An estimate of the time scale for transit and mixing of the gases, and impact on detector response would be useful. Current zone models also do not account for the effect of structural obstructions on the ceiling layer and its potential convective cooling.

Suppression system effectiveness would be affected by the water droplets hitting an obstruction, leaving a hole in the spray pattern. If more than one sprinkler were activated, the hole in the spray pattern might be somewhat negated. It is well known that sprinklers cannot put out a fire that is burning below a low barrier.

4.5.2.3 Threshold for Thermal Equipment Damage Criteria

Failures of equipment exposed to the harsh environment of a fire and the subsequent suppression activities are typically modeled by a threshold value of an appropriate parameter. This threshold value is referred to as the "equipment damage criterion." As an example, a threshold surface temperature is usually considered as a damage criterion for cables. Relative humidity and smoke concentration may be more suitably considered for small electrical equipment such as relays, but are not considered in current PRAs.

Establishing damage criteria is a complex process. Equipment exposed to the thermal environment of a fire may fail either temporarily or permanently. As an example, an electronic circuit may temporarily fail (not respond or respond incorrectly) when exposed to high temperature; however, it may recover performance when the temperature drops. The failure criteria for equipment are also dependent on equipment function. As an example, small insulation leakage current can cause failure of an instrument cable, whereas, the same amount of leakage in low-voltage power cable could be inconsequential.

Owing to these difficulties, among others, the damage criteria typically used in PRAs are uncertain. This uncertainty directly affects the fire PRA results, since the damage criteria are used to determine the time available for successful suppression.

4.5.2.4 Effect of Smoke on Equipment

Smoke from a fire that starts in one zone might propagate to other zones and potentially damage additional equipment. Currently, fire PRAs do not treat the question of smoke propagation to other areas and their effect on component operability in a comprehensive manner. The extent to which the issue is addressed depends on the analyst and, if it is addressed, it is typically addressed qualitatively.

The current general understanding on this issue is described below:

(1)   Smoke, depending on what is in it (such as HCl from burning polyvinyl chloride (PVC) insulation), causes corrosion after some time. A little smoke has been shown to cause damage days later if the relative humidity is 70 percent or higher. Navy experience has shown that corrosion can be avoided if the equipment affected by smoke is cleaned by a forceful stream of water containing non-ionic detergent, and then rinsed with distilled water and dried.

(2) Smoke can damage electronic equipment (NUREG/CR-6476), especially computer boards and power supplies. Fans cooling the electronic equipment can introduce smoke into the housing, strongly affecting the extent of the damage.

(3) Smoke can also impair the operation of relays in the relay cabinet by depositing smoke products on the contact points. Again, the forced cooling of the relay panel can exacerbate the situation.

4.5.2.5 Operator Actions

Because of the state of the art of human reliability analysis (HRA), large uncertainties are generally associated with the probability of the success of operator actions. For fire events, the modeling of operator actions becomes more complex because of the necessity to account for the effects of the fire and smoke on operators.

4.6 CONCLUSION

Since Appendix R was issued in 1980, the probabilistic risk assessment methodology has been developed and used over the last 15 years by the NRC and the U.S. nuclear industry to

(1) determine plant risk from fire events as part of general assessments of the total risk profile from plant operations; and

(2) identify vulnerabilities to fire events and implement cost-effective safety improvements to either eliminate or reduce the impact of these fire vulnerabilities.

To date, PRA methods have not been used to implement current fire protection regulations.

A review of 12 PRA studies conducted by the NRC, EPRI, and nuclear utilities to assess plant risk, including risk from fire events, yielded the following observations:

- Given the same plant configuration and parameters, the absolute results of fire PRAs vary significantly because of the data,

methods, and assumptions used (particularly between those sponsored by NRC and EPRI);

- Given similar data, methods, and assumptions, there are major differences in estimated fire CDF that can be explained by plant-specific system design and the embedded level of redundancies in safety functions;

- Most studies indicate that the majority (in some cases as much as 90 percent) of the risk from fires in nuclear power plants comes generally from three or four fire areas, such as the control room, cable spreading room, and the switchgear room.

- Fire protection analysis using PRA differs in many respects to analysis per NRC requirements in Appendix R. For example, even though most fire PRAs have identified fires in the control room and the cable spreading room as significant contributors to core-melt probability, a coincident loss of offsite power is not included in the scenarios. This is quite different from the requirements of Appendix R, which requires an assumption that offsite power is lost coincident with a fire in the control room. The significance of a control room fire as modeled in PRAs is usually attributable to scenarios other than the loss of offsite power (e.g., a control room fire in a PWR may, among other things, cause the power-operated relief valves (PORVs) to open spuriously).

A preliminary conclusion has been reached by the NRC staff that the fire PRA and FIVE methods have been successfully used to achieve the objectives of the IPEEE regulatory program to identify plant vulnerabilities to fire events and implement cost-effective safety improvements to either eliminate or reduce the impact of these fire vulnerabilities. The fire IPEEE conducted by the Quad Cities nuclear power station has been cited by the NRC staff as an example of the success of the IPEEE program and an example of the use of fire PRA and/or the FIVE methods to identify vulnerabilities not addressed by Appendix R.

Various uncertainty issues that have been stated to be associated with fire PRA and modeling are discussed in this chapter. A number of different areas of a fire protection program can be analyzed without the need for fire modeling (e.g., fire protection equipment surveillance and maintenance test intervals). For these cases, the issue of uncertainty can be formally addressed and incorporated in the decisionmaking process. This is discussed further in Chapter 6.

In other cases in which evaluation of the issue necessitates the use of fire modeling, the portion of fire modeling that predicts the fire heat-release rate was differentiated from the portion that predicts the thermal environment. Larger uncertainty ranges are associated with the predicted heat-release rate than with the thermal environment. In any case, the heat-release rate of the fire source, knowing the current state of the art, may be best estimated conservatively by using simplified engineering evaluation, subjective judgment, and extrapolation of actual fire events or fire tests. A critical analysis of trial applications to assess the usefulness of results and insights that may be gained from fire PRA and modeling methods for improving regulatory decisionmaking in light of the uncertainties discussed in this chapter is presented in Chapter 6.

5 DEVELOPMENTS AND PRACTICES OUTSIDE NRC AND U.S. NUCLEAR INDUSTRY

5.1 DEVELOPMENTS IN NUCLEAR INDUSTRY IN FRANCE

The main measures concerning fire protection in French nuclear power plants are (1) fire prevention by physical separation between redundant safety trains, fire confinement, and protection for cables; (2) fire protection by zones and alarms transmitted to the control rooms;(3) fire fighting, including escape paths, containment of smoke, and suppression systems. These requirements are defined in a document called RCC1 (rules for fire protection in pressurized-water reactors (PWRs)). The Directorate for the Safety of Nuclear Installations (DSIN) of the Ministry of Commerce issues the basic safety requirements and is supported by the Institute of Protection and Nuclear Safety (IPSN) of the French Atomic Energy Commission (CEA) in the review of the designs.

The safety roles are divided between the regulator and operator as follows:

- Government authorities determine the safety objectives.

- Plant operators propose the means to meet the objectives which is reviewed by the safety authority and approved, if satisfactory.

The authorities exercise oversight in the design stage, the construction stage, and the operation stage.

Once the installation is in operation, French safety authorities analyze the causes and consequences of minor fires and also inspect the facilities. Because of the standardization of French PWRs, the discovery of an anomaly in one unit leads to a corresponding modification of all the PWR units concerned. The efficacy and rapidity of the interventions in the event of fire are an essential complement to the protective measures. For example, each nuclear power plant draws up a number of fire action sheets that define, for each facility or fire zone and for each staff category, the existing means of detection and intervention and the types of action to be carried out in order to limit the fire and its consequences. DSIN takes measures to ensure that this quality (of training and operation) is maintained at a satisfactory level throughout the lifetime of the installation.

The following is description of computer codes that have been developed by the French nuclear industry. The material is purely descriptive and no attempt has been made to provide a critical analysis since the detailed documentation for the code was not reviewed, and the authors did not use the code for examining any specific problems.

5.1.1 The FLAMME-S Fire Computer Code

Although limited efforts were underway earlier to use fire models, IPSN initiated an intensive effort in 1993 to develop the FLAMME code to quantify the thermal response to the environment and equipment and use the results of this analysis in their fire PRAs . The objective is to predict the damage time for various safety-related equipment. The FLAMME-S version (Bertrand et al., 1996) may simulate the development of fire in one of several rooms in a parallelopedic form with vertical or horizontal openings, confined or ventilated, containing several targets and several combustible materials. The design of the code is based on the assumption of a three-zone model: (1) cold zone, (2) hot zone, and (3) plume of flame. In particular, the code can quantify the thermal response to equipment located in the plume directly above the fire source, in the hot gas layer outside the plume, or next to the fire source exposed to heating by thermal radiation. The code not only calculates the fire within a compartment, but also the consequences of the potential extension to other compartments through communicating elements. A compartment can also contain several cable trays as well as several pieces of equipment such as electrical cabinets. The code has several additional features: (1) it can use liquid or solid fuels and a mixture of flammable product gases in each compartment; (2) ventilation can be forced or natural; (3) a fire damper or door can be closed gradually, as a function of time or of temperature or of heat flux; (4) there can be mass exchange between the plume, upper layer, lower layer, and the ventilation; and (5) the fire can be on the floor or above a cabinet. A pictorial representation of the FLAMME-S code is shown in Figure 5.1.

Some additional characteristics of the fire model are that oxygen needed for combustion is taken from the compartment with the fire in it, and the flame is a point source of radiant flux. Calculated target temperatures are surface and average inside temperatures, and cable trays can ignite grid by grid. Equipment can be impaired by the temperature or the heat flux, and impairment depends on the location of the target. The shape of the fire plume need not be a V, which permits realistic radiation view factors, and it is possible to handle the partial shielding of the thermal radiation by an intervening object, but this is difficult, so it is

seldom done.

The code can be used to predict the following phenomena: (1) oxygen concentration and combustion products; (2) pressure; (3) mass rates of gas inflow and outflow via openings; (4) temperatures of the gases (bottom zone, top zone, plume, in equipment), of the walls, and of any equipment in the compartment; (5) heat flux emitted by the flame and incident heat flux in walls and on installed equipment surface; and (6) damage (functional impairment or combustion) and time to damage of equipment.

The code is subject to a specific quality assurance procedure. This quality assurance procedure requires, in particular, the production of a "life history" for the code, known as the "qualification dossier" containing the following main documents: (1) functional specifications (design principles of the code); (2) technical specifications; (3) physical and mathematical models (formulas used in the development phase); (4) numerical descriptions (approximations of the formulas used); (5)

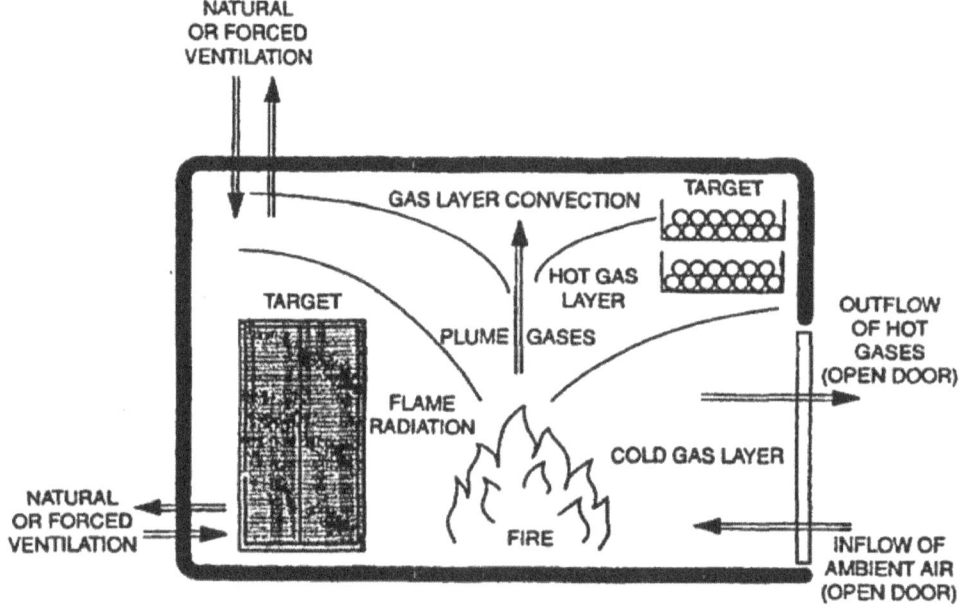

Figure 5.1
Quantification of Thermal Response by FLAMME-S Code (Three Zone Model)

computer description (code and its computer environment); (6) qualification report (comparison of tests and calculations); (7) data sensitivity study (data uncertainties and numerical sensitivity); and (8) a user's guide containing information on adequate data selection and examples. The last three elements are considered the most important for ensuring quality. Any modifications or upgrades to the code leads to the issuance of a new version, and the quality assurance process is applied before the upgrade is released for use.

The first version of FLAMME-S is operational. It will be used for the fire PRA study for the 900-MWe PWRs being conducted by IPSN and for other safety assessments of issues uncovered during plant inspections (some safe-shutdown equipment was found to be outside protected areas).

5.1.2 The MAGIC Fire Computer Code

The French utility, Electricité de France (EdF), uses a different computer program, called "MAGIC." MAGIC is a multicompartment zone model, and it is used by safety engineers at EdF as a basis for discussions of fire safety provisions. The lower layer stays cold, and the heat transfer through the walls is one-dimensional conduction, with the heat going into the next compartment. There can be several (up to about 5 or 6) fires in a compartment, each with a separate plume. Radiation can be calculated between the flame, walls, and gases; gases are treated as semi-transparent, and the walls as "gray." The fire can be limited by lack of oxygen, in which case the unburned gas in the next compartment flames. Research work for MAGIC is carried out both by the French Government research agency, Centre National de la Recherche Scientifique (CNRS) at Poitiers, and by international cooperation.

5.1.3 Fire Computer Code Validation

FLAMME-S is being validated by fire tests in two IPSN laboratories, Grenoble and Cadarache, which have chambers of volumes from 5 m³ to 3600 m³ for use in small, medium, and large-scale tests in various configurations. Tests are conducted to determine input parameters for fire

and ventilation computer codes for various configurations and fuels commonly used in nuclear installations, validation of fire and ventilation computer codes for various configurations, and obtaining a better understanding of fire phenomena. The experiments include fire propagation from the first ignition source to other combustible materials in the same compartment, fire propagation from the originating compartment to other compartments through communicating elements, and defining real cable dysfunction phenomena. At present, 40 intermediate and large-scale tests have been performed in various configurations with fuels common in nuclear facilities and used in the code validation and qualification process. A program of experiments, which comes under the IPSN's "Five Year Plan," is currently in process in order to gain a better understanding of the fire phenomena, increase confidence in the results of the digital simulation, and, particularly, to qualify new versions of the code.

IPSN has conducted a study (NRC Translation Numbers 3383 and 3384) to evalate the capabilities of the COMPBRN code including a comparison of code output with experimental results.

The MAGIC code has been qualified with real-size experiments. A selection of real-size tests from the literature of several countries, including the U.S., is used for a direct comparison with code results. EdF also plans to compare MAGIC with CFAST, COMPBRN, and FIVE tables through a memorandum of understanding between EdF and the Electric Power Research Institute (EPRI).

IPSN and EdF are involved in several international collaborative programs to improve its code and capabilities in this area. One current joint activity by IPSN and EdF is to compare MAGIC and FLAMME calculations for a 10-m × 5-m × 3-m concrete room with a pool fire at the center of the floor surface, and with three targets, at different elevations. They will compare calculated temperatures, layer height, pressure, oxygen concentrations, wall temperatures, and the concentrations of other species.

5.1.4 Conclusion

In conclusion, IPSN and the utility Electricité de France have considerable efforts underway for developing and utilizing fire PRAs supported by the fire computer codes discussed above. They have concluded that this tool provides useful information for safety assessments to supplement deterministic analysis on which reactor design and fire protection provisions are based. The fire PRA will identify the most significant locations where vulnerabilities exist. The results will be used to support the necessary analysis within the framework of the periodic safety assessments conducted every ten years in France for each plant.

5.2 DEVELOPMENTS IN U.S. AND FOREIGN BUILDING INDUSTRIES

The building industry in the Unites States and several other countries (e.g., Japan, Sweden, Finland, Australia, Canada, United Kingdom, and New Zealand) has moved toward adopting performance-based codes (note "code" here denotes a regulation). Among the benefits identified are designs to achieve fire safety that are better and less expensive than those achieved with prescriptive code provisions. The Japanese Ministry of Construction is in the forefront of these efforts. The initiatives in the United States, Japan, and United Kingdom are summarized below to illustrate the nature of and progress in developing performance-based fire safety codes. Although the main goal of fire protection for commercial buildings, that is, life safety, is different from that for nuclear power plants, the information in this section is presented because several features of the fire models and computer codes for the two applications would be similar. Also, other important goals in building fire safety are the assessments of the fire endurance of walls and floors to determine fire fighting capability, and spread of fire to nearby structures, both of which are applicable to nuclear power plants. Appendix A describes the initiatives in New Zealand, Australia, Canada, and Nordic countries, and provides further details of the Japanese initiative.

5.2.1 United States

Many players, both private and public, are involved in the development of fire safety codes in the United States. Model code organizations (private) develop the basic code requirements, which are then adapted and adopted by numerous legislative bodies at the State and local levels. One common feature in the U.S. codes is the "equivalency clauses," which allow for the acceptance of alternative approaches that meet the intent of the prescriptive requirements and which are intended to allow flexibility and foster innovation.

Initial deviations from prescriptive requirements were substantiated in the form of logical arguments, data from tests, or example (it was accepted elsewhere and has worked). Recently, engineering models and calculations are being submitted to support deviations from prescriptive provisions. With positive experiences, code officials are becoming more comfortable with calculations for egress and fire growth in granting variances. It has been recognized that performance codes are a worthy goal in that they promise to allow safety to be maintained, while improving design flexibility and reducing cost. Although a more formal equivalency-determination system has been introduced in some areas (e.g., Health Care Occupancy chapter of National Fire Protection Association's Life Safety Code), it has been recognized that the move toward performance-based fire safety codes will require fundamental changes in fire safety regulation.

Most of the prescriptive building codes used by the various State and local governments in the U.S. are derived from one of three model codes. Currently, the three model code organizations are working to create a single prescriptive "International Building Code" (Traw, 1998), which is scheduled for release in the year 2000. This "international" code will be a selective combination of prescriptive requirements from the existing three model codes. In a parallel effort, this same group is developing a performance-based version of the "International Building Code" to be called the "International Performance

Code." The target date for completing the performance version is early 2000.

In July 1995, the National Fire Protection Association published a document titled *NFPA's Future in Performance-Based Codes and Standards—Report of the NFPA In-House Task Group* (NFPA, 1995). This document established NFPA as a participant in the performance based code arena. Using the guidance contained in the document and support from the in-house task group, several NFPA technical committees are pursuing the conversion of their respect code or standard from prescriptive to performance. Currently, the two most active committees in this area are Safety to Life and Atomic Energy.

Within the past few years, the Society of Fire Protection Engineers has initiated a number of efforts aimed at providing the engineering support necessary for implementation of a performance-based code system. A fundamental activity in this area is the development of engineering practice documents. These documents are intended to provide peer-reviewed guidance concerning appropriate processes and practices for conducting a performance-based design. Specific initiatives include establishment of several engineering task groups to address issues such as fire model evaluation, manuals of practice, building code liaison, design team liaison, and performance (Custer and Meacham, 1997) In addition, SFPE is providing educational support by conducting seminars, symposia, and short courses. The SFPE continues to publish technical guidance such as the *SFPE Handbook of Fire Protection Engineering* and the *SFPE Journal of Fire Protection Engineering* (Meacham, 1996).

Prediction tools are slowly gaining acceptance within the regulatory community, particularly for simpler problems, where experts can judge if the predictions are reasonable. However, for more complex problems there is difficulty in understanding the uncertainties in a calculation. The National Institute of Standards and Technology (NIST) has proposed to relate the predictive uncertainty—including both the calculational uncertainty and the uncertainty in the impact data as it propagates through the calculation—to a design safety factor that will

ensure that an undesirable result will not occur. NIST is implementing an effort to develop the scientific understanding and calculational models to allow the adoption of performance-based fire safety engineering.

NIST (Snell et al., 1993) has proposed a three-level fire safety engineering framework, in which "framework" is defined as a conceptual scheme, structure, or system. The first level is primarily analytical, containing calculational methods for determining fire risks and benefits. The second level, largely phenomenological, has tools for predicting fires and for measuring the performance of fire safety technologies or actions. The third level involves the knowledge, measurement methods, and data needed to support the tools.

The General Services Administration (GSA) uses a collection of fire models, FPETOOL, to evaluate the fire safety of the Government-owned or -leased buildings in its inventory (Stroup, 1993). For each occupancy, a number of design fires, those that would cause the most severe impacts on the building and its occupants, are assumed. The fire scenarios are modeled to determine the effects on life safety, property, and mission. Finally, the model (or models) is used to evaluate the effect of various protection schemes on the identified fire safety risks. The GSA funds research necessary for the further development of FPETOOL and has funded instruction at NIST for GSA personnel.

Training in the use of FPETOOL, CFAST, and other computer models is provided today in fire protection engineering courses at the University of Maryland, Worcester Polytechnic Institute, and other educational institutions. Many fire protection engineering firms employ personnel who are expert in the use of these models.

Credibility of the prediction tools as an equivalency method is still developing among regulators in the U.S. building industry. The need for specific models or calculational methods to be reviewed and sanctioned by independent bodies has been recognized as necessary to advance the adoption of performance-based requirements. Manuals of practice that lay out the proper procedures (e.g., data sources, appropriateness of a model relative to its assumptions, the role of sensitivity analysis, accuracy, and uncertainty estimates) are being developed.

The following are descriptions of computer codes that have been developed and used in the U.S. building industry for performance-based fire protection analysis. The material presented for the codes is purely descriptive and no attempt has been made to provide a critical analysis since the detailed documentation for the codes were not reviewed, and the codes (except for CFAST) were not run for any specific nuclear power plant problems. The CFAST code was run for a specific nuclear power plant issue, along with the COMPBRN code, and the results including a critical analysis are presented in Chapter 6.

5.2.1.1 The Program FPETOOL

FPETOOL is a collection of computer-simulated procedures providing numerical engineering calculations of fire phenomena to the building designer, code enforcer, fire protection engineer and fire safety-related practitioner. The latest version incorporates an estimate of smoke conditions developing within a room receiving steady-state smoke leakage from an adjacent space. Estimates of human viability resulting from exposure to developing conditions within the room are calculated on the basis of the smoke temperature and toxicity. There is no modeling of human behavior. An estimation of the reduction in fire heat-release rate due to sprinkler suppression is also included in the latest version.

FPETOOL (Deal, 1995) is a compilation of several modules grouped into six categories. These categories are

- SYSTEM SETUP
- FIREFORM
- MAKEFIRE
- FIRE SIMULATOR
- CORRIDOR
- 3rd ROOM

SYSTEM SETUP is a utility routine. It allows the user to change file destination and source directories, change operating units, and change

screen colors. The menu choices are presented above as they appear to the user.

FIREFORM is a collection of quick procedures designed to solve primarily single-dimensional questions. Such questions might be: How hot is the ceiling jet 3 m (~10 ft) from the center of plume impingement? How long will it take for 50 people to evacuate from the 7th floor of this building to ground level? When will this fuel item exposed to the fire source ignite?

MAKEFIRE is a collection of routines for creating fire files. These files have three columns of data: time, fire heat-release rate, and fuel pyrolysis rate. The user has the option of letting the program determine when the second item ignites, defining a fire according to a generic "t-squared" formula,* or describing another specifically applicable heat-release rate curve.

FIRE SIMULATOR is a procedure that can predict the thermal environment from a fire using a one-room, two-zone, two-vent model with capability to predict fire detection and sprinkler actuation..

CORRIDOR is a procedure that predicts the characteristics of a moving smoke (hot gas) front in a corridor. The procedure is formulated for spaces with large length-to-width ratios.

3rd ROOM is procedure that predicts smoke conditions (toxicity and visibility) developing in a room and the subsequent reduction in visibility and threat to human life.

The last three modules may be used sequentially. FIRE SIMULATOR predicts fire-generated effects within the room of origin. Smoke outflow from FIRE SIMULATOR may be used as smoke inflow to the CORRIDOR module. Smoke conditions predicted with the CORRIDOR module can be used to define conditions on the "fire side" of the door to the 3rd ROOM.

---

*This is one type of power-law growth for heat generation modeling.

5.2.1.2 The Program CFAST

CFAST (Peacock et al., 1993b and 1997) is a multi-room zone model with comprehensive capabilities. Some of its features are described briefly in the sections that follow

5.2.1.2.1 Fires

Within CFAST, a fire is a source of fuel that is released at a specified rate. This fuel release rate is then converted into enthalpy (the conversion factor is the heat of combustion) and mass (the conversion factor is the yield of a particular species as it burns). Burning can take place in the portion of the plume in the lower layer (if any), in the upper layer, or in a door jet. For an unconstrained fire, all of the burning will take place within the fire plume. For a constrained fire, burning will take place where there is sufficient oxygen. If insufficient oxygen is entrained into the fire plume, unburned fuel will successively move into and burn in the upper layer of the fire room, the plume in the doorway to the next room, the upper layer of the next room, the plume in the doorway to the third room, and so forth, until it is consumed or exhausted outside.

The latest version of CFAST has the capability to independently track several fires in one or more rooms of the building. These fires are treated as totally separate entities, that is, with no interaction of the plumes or radiative exchange between fires in a room.

Like most current zone fire models, this version of CFAST does not contain a pyrolysis model to predict fire growth. Rather, pyrolysis rates for each fire modeled define the fire history. The similarity of that input to the real fire problem of interest will determine the accuracy of the resulting calculation. The user must account for any interaction between the fire and the pyrolysis rate. Future research should remove this limitation.

5.2.1.2.2 Plumes and Layers

Above any burning object, a plume is formed that is not considered to be a part of either layer, but that acts as a pump for enthalpy and mass from the lower layer into the upper layer (upward only). For the fire plume, CFAST does not use a point source approximation, but rather uses an empirical correlation to determine the amount of mass moved between layers by the plume.

Two sources exist for enthalpy and mass transport between the layers, within and between rooms. Within the room, the fire plume provides one source. The other source of mixing between the layers occurs at vents, such as doors or windows. Here, there is mixing at the boundary of the opposing flows moving into and out of the room. The degree of mixing is based on an empirically derived mixing relation. Both the outflow and inflow entrain air from the surrounding layers. The flow at vents is also modeled as a plume (called the door plume or jet), and the same equations as those for the fire plume are used, with two differences. First, an offset is calculated to account for entrainment within the doorway; second, the equations are modified to account for the rectangular geometry of vents compared to the round geometry of fire plumes. All plumes within the simulation entrain air from their surroundings according to an empirically derived entrainment relation. Entrainment of relatively cool, non-smoke-laden air adds oxygen to the plume and allows the fuel to burn. It also causes the plume to expand in the shape of an inverted cone as it moves upward. The entrainment in a vent is caused by bidirectional flow and results from a phenomenon called the "Kelvin-Helmholz instability." It is not exactly the same as a normal plume, so some error arises when this entrainment is approximated by a normal plume entrainment algorithm.

5.2.1.2.3 Vent Flow

Two kinds of flow come through vents. The first is referred to as "horizontal flow." It is the flow that is normally thought of in discussing fires. It encompasses flow through doors, windows, and so on. The other is "vertical flow," and it can occur if there is a hole in the ceiling or floor of a compartment. This latter phenomenon is particularly important in three disparate cases: on a ship, in the role of fire fighters engaged in roof venting, and fire propagation in typical containments for nuclear power plants.

Flow through normal vents is governed by the pressure difference across a vent. Two situations give rise to flow through vents. In the first situation—usually thought of in fire problems—air or smoke escapes from a compartment by buoyancy. The second type of flow is due to expansion that is particularly important when conditions in the fire environment are changing rapidly. Rather than depending entirely on density differences between the two gases, the flow is forced by volumetric expansion. The earlier version of this model did not solve this part of the problem entirely correctly. In most cases, the differences are small, except for rapidly changing situations. However, these small differences become very important in a situation in which flows are due to small pressure differences, such as will occur with a mechanical ventilation system. Atmospheric pressure is about 100,000 pascals (Pa), and fires produce pressure changes from 1 to 1000 Pa; to solve these interactions correctly, we must be able to follow pressure differences of ≈ 0.1 Pa out of 100,000 Pa for the overall problem, or 1E-4 for adjacent compartments.

5.2.1.2.4 Heat Transfer

Heat transfer is the mechanism by which the gas layers exchange energy with their surroundings. Convective transfer occurs from the layer to the room surfaces. The enthalpy thus transferred in the simulations conducts through the wall, ceiling, or floor in the direction perpendicular to the surface only. CFAST is more advanced than most models because it allows different material properties to be used for the ceiling, floor, and walls of each room (but all the walls of a room must be made of the same material). Additionally, CFAST uniquely allows each surface to be composed of up to three distinct layers, which are treated separately in the conduction calculation. This not only produces more accurate results, but allows the user to deal naturally with the actual building construction.

Radiative transfer occurs among the fire(s), gas layers, and compartment surfaces (ceiling, walls, and floor). This transfer is a function of the temperature differences and the emissivity of the gas layers as well as the compartment surfaces.

For the fire and typical surfaces, emissivity values only vary over a small range. For the gas layers, however, the emissivity is a function of the concentration of species that are strong radiators: predominantly smoke particulate, carbon dioxide, and water. Thus, errors in the species concentrations can give rise to errors in the distribution of enthalpy among the layers, which results in errors in temperatures and, consequently, errors in the flows.

5.2.1.2.5 Species Concentration and Deposition

When the layers are initialized at the start of the simulation, they are set to ambient conditions. These conditions are the initial temperatures specified by the user, and 23 percent by mass (20.8 percent by volume) oxygen, 77 percent by mass (79 percent by volume) nitrogen, a mass concentration of water specified by the user as a relative humidity, and a zero concentration of all other species. As fuel is pyrolyzed, the various species are produced in direct relation to the mass of fuel burned (this relation is the species yield specified by the user for the fuel burning). Since oxygen is consumed rather than produced by the burning, the "yield" of oxygen is negative and is set internally to correspond to the amount of oxygen needed to burn the fuel. Also, hydrogen cyanide and hydrogen chloride are assumed to be products of pyrolysis, whereas carbon dioxide, carbon monoxide, water, and soot are products of combustion.

Each unit mass of a species produced is carried in the flow to the various rooms and accumulates in the layers. The model keeps track of the mass of each species in each layer and knows the volume of each layer as a function of time. The mass divided by the volume is the mass concentration, which, along with the molecular weight, gives the concentration in volume percent or parts per million, as appropriate.

The species concentrations are important in that they can be used to calculate the toxic impact of the gases on persons trying to escape from the fire. This calculation, along with others, is carried out in a set of programs called HAZARD I, in which CFAST is embedded.

5.2.1.2.6 Code Validation

NIST has tried to improve CFAST so that when it is used within the range of variables for which it has been verified, dependable results are obtained. The CFAST model has been subjected to a wide range of comparisons to experimental data. The comparisons range from simple single-compartment fires (Deal, 1990), multi compartments on a single floor and a seven story hotel (Peacock, et al., 1993a), to large aircraft hangers (Duong, 1990, Davis et al., 1996b). For variables deemed of interest to the user of the model, the CFAST model provided predictions of the magnitude and trends (time to critical conditions and general curve shape) for the experiments, which range in quality from a few percent to a factor of 2 to 3 of the measured values.

5.2.1.3 The Program FASTLite

FASTLite is a one-room to three-room version of the program CFAST, packaged with most of the FIREFORM routines of FPETOOL on a CD-ROM disk. Both CFAST and FPETOOL are described above. The FASTLite outputs can be printed as tables or to a spreadsheet, and as graphs of temperatures, layer heights, and burning rates vs. time.

This program is available as "FASTLite," Special Publication 899 from the U.S. Department of Commerce, NIST, Fire Modeling and Applications Group, Gaithersburg, Maryland 20899.

5.2.1.4 Codes for Simulating Smoke Travel During Fires

Mathematical models are currently being developed to calculate smoke travel from fire that may occur in a non-critical area of the power plant to other areas in which operators and others must perform their duties. This problem, of course, is of major concern for life safety in building fires, as well as in nuclear power plant fires, so a large degree of interest has resulted in well-developed and validated models.

The best known model for smoke travel between interconnecting rooms is ASCOS, which is described in the ASHRAE (American Society of Heating, Refrigeration and Air Conditioning Engineers) publication "Design of Smoke Management Systems," Atlanta, Georgia (1993). However, the input routine for this model is somewhat tedious and it is easy to make errors. An improved program, with a graphical input routine, designed for personal computers (PCs), is CONTAM. CONTAM (Walton, 1994) lets the user draw the connections between rooms and between rooms and the outside on a sketch of the floor plan (which need not be drawn to scale), and enters these to the calculational software. Also, leakages between rooms, or to the ceiling or the floor, can be entered. Species such as acid gases, as well as vision-impeding soot, are carried by the "smoke." The user must specify the smoke output from the fire, as well as the heat output, as a function of time.

In the case of a single room, CFAST can be used to estimate the smoke content of the upper layer. In a tall single room, such as an atrium, the results will be more accepted if the "Heskestad" plume model (used by the NFPA 92B Guide) is used instead of the "Zukowski" plume model normally used in CFAST (Zukowski er al., 1980/1981). The amount of material entrained in the plume could be in error for these tall plumes by a factor of 2 if the room is more than 10 m or so high.

5.2.1.5 Computational Fluid Dynamics (Field) Models

The application of the techniques of computational fluid dynamics (CFD, also called "field modeling") to fire problems has been rapidly increasing during the past decade since the application of this technique to the deadly King's Cross underground station fire in London (Simcox et al., 1989). This application provided insights on the observed fire growth that could not be drawn without the analysis or zone models. This method of modeling smoke and heat flow requires that the region of interest be divided into a collection of small rectangular boxes, or control volumes. Zone models use only two or three control volumes per room; a CFD model may have 100,000 or more control volumes.

Heat is released in several control volumes over time. The resulting flow (or exchange of mass, momentum, and energy) between control volumes is determined so that these three quantities are conserved. Momentum conservation equations are determined by the Navier-Stokes equations.

These fluid flow equations are expressed mathematically as a set of simultaneous non-linear partial differential equations, and after some manipulation are solved each time step for each of the control volumes. Turbulent flow problems may require the solution of additional equations. Obviously, considerable computer capability is required both to run the calculations and to display the results.

Four current CFD models used for fire problems are: FLOW-3D (British Harwell Laboratory); JASMINE (British Fire Research Station); LES (NIST Building and Fire Research); and KAMELEON (Norwegian SINTEF NBL and Sandia National Laboratory).

The application of CFD to fire problems opens up the possibility of modeling smoke and heat flow around obstructions and in complex geometries. The impact of forced ventilation or wind on smoke flow can easily be modeled. Recent improvements have also allowed radiation exchanges between the fire and the surroundings to be included. In some instances, simple chemistry can be included in CFD calculations, but the scale size of the reaction region and the present speed of computers prevent the implementation of these calculations in room-size fires. Other fire-related problems that can be included are the activation of heat and smoke detectors and the penetration of water sprays through the fire plume.

A lack of appropriate validation studies hinders assessing the accuracy of field model calculations. Currently, a study by NIST and the National Fire Protection Research Foundation is underway to investigate the use of the Large Eddy Simulation (LES) model to analyze the interaction of sprinklers, vents, and draft curtains. As part of this effort, a number of full-scale tests are being conducted to develop data for verification of the model results. Among other things, this modeling effort involves calculating fire growth, time of activation for multiple sprinklers, and the impact of the sprinkler spray on the burning commodity (McGrattan et al., 1997a). Another effort, being conducted by NIST with sponsorship from the Mineral Management Service, is aimed at verifying the use of a version of the LES model to predict the development and spread of smoke plumes from burning oil spills in the outdoors (McGrattan et al., 1997b). As the use of field models becomes more widespread, additional verification efforts will no doubt be conducted.

A still more accurate way to calculate the smoke content at a given place in the room versus time would be to use a "field" model, such as FLOW-3D or LES, which divides the room into thousands of zones. Typically, field models require computer workstations. However, NIST is investigating the possibility of running field models on PCs to solve specific fire protection problems (Walton, 1996).

5.2.2 Japan

The Japanese have a significant initiative for developing fire models. Beginning a decade ago, they developed a detailed methodology that can be used to establish equivalency to the Building Standard Law of Japan. This method was developed in 1988 (Wakamatsu, 1989) and has been growing in use since. The number of "Article 88 Appraisals" has increased to hundreds per year, although they are still limited to special projects with unique requirements that could not be easily achieved under the prescriptive law.

The Japanese are able to accomplish this because they have a single, national code promulgated by the Ministry of Construction (MOC) that is enforced locally. The code allows equivalency, as do the U.S. codes, but the determination of equivalency rests with the MOC. Thus, when the Building Research Institute (part of MOC) published the calculational method, it represented a "sanctioned method" for establishing equivalency. Further, a mechanism has been established whereby the local authority can solicit the advice of MOC on the appropriateness of a calculation, further adding to the comfort of the authority having jurisdiction.

Published in four volumes, the method represents a manual of practice for evaluating the fire safety of a building. Volume 1 discusses the goals and objectives of achieving safety and presents several case studies as examples. Volume 2 covers fire prevention and containment. Calculation methods for predicting fire and smoke spread within a building are included along with typical data needed to perform the calculations for most buildings. An example calculation for an atrium is included. In Volume 3, egress and tenability calculations are covered. Necessary data, including occupant characteristics and loadings by occupancy type, are given along with several example calculations. The fourth volume is a manual of fire-resistant design containing design standards, calculation methods, and examples. For common assemblies, charts and simplified calculations are presented.

Although the Japanese have no performance code, they do have a performance-based method that is officially sanctioned as providing equivalent designs. They have a manual of practice that gives details of the calculation methods and all necessary data, along with numerous examples. They have also established a system by which local authorities can receive assistance in evaluating the appropriateness of the calculation in case they feel uncertain or uncomfortable in making that decision.

The Japanese have now initiated the second phase of their program to completely evolve to the performance-based building fire regulation system to replace the current prescriptive law (Nakaya, 1998). A research project on the development of assessment methods of fire safety performance (the level of safety that must be reached by each requirement of the performance code) is ongoing.

5.2.3 United Kingdom

In principle, the United Kingdom moved to a performance-based model building code by adopting the Housing and Building Control Act of 1984 (United Kingdom, 1985). This system replaced prescriptive requirements with broad functional statements. The basic regulation was then supplemented by a series of "approved documents." These documents spell out a way by

which the intent of the regulation can be deemed to be satisfied. It was understood that these approved documents would then, in the long term, constitute fire safety engineering guidelines. This was seen as requiring a long time and significant funding to accomplish. Thus, the first edition of the approved documents consisted, essentially, of a republished old prescriptive code. Compliance with the old code, therefore, was deemed as compliance with the new regulation as well. Other designs could be offered, however, if they were approved by the local building authority. Recently, approval authority for performance-based designs in housing construction has been removed from the local authorities and vested in "approved inspectors" (Rackliffe, 1998). Enforcement remains the domain of the local authorities. The approved inspectors are private firms or individuals who are paid as outside experts by the builder. They have the expertise to judge the value of the design. Local authorities cannot appeal the decisions rendered by the approved inspectors.

The first step toward putting some flesh on these performance "bones" was a study by H. L. Malhotra (1987) (then recently retired from the Fire Research Station) commissioned by the Department of Environment. Malhotra considered that the building fire safety objectives were

- life safety
- prevention of conflagration
- property protection

This particular tripartite split is notably very general. "Life safety" is so general as to be nearly akin to "public welfare." Prevention of conflagration is certainly important and essential, yet some quite unrelated issues are placed together, that is, building construction, lot sizes and zoning, and fire fighting operations. Finally, some people disagree that property protection, apart from conflagration control, is a government function.

To develop further details in his plan, Malhotra examines, in his study, several building codes from different parts of the world and proposes a model scheme for occupancy classifications. In general, this scheme is very similar to the one used in the Uniform Building Code and other traditional codes. Classifications are given for such occupancies as residential, education, business, and factory. Note, however, that the traditional concepts of regulation according to occupancy type are not founded on sound engineering principles. A framework based on fire safety engineering concepts would demand that such "top-level" classifications be based on (1) degrees of hazard, (2) degrees of risk, or (3) similarity of fire environments. The traditional occupancy classifications are simply based on *uninformed judgment*, that is, judgment not supported by physics, statistics, or even case-trend analysis. The most essential objectives of a rational, performance-based building code should be to present the scientific bases for a "top-level" building categorization scheme.

Malhotra's scheme includes major engineering modules for

*   the design of means of escape
*   fire development within the initial space of fire origin
*   fire propagation from room to room
*   fire propagation from the burning building to another building
*   detection, fire fighting, and extinguishment
*   fire safety management (e.g., staffing, training, maintenance of equipment)

These more detailed building blocks are developed in some detail in Malhotra's study. Although conceptual planning of the principles of fire protection has progressed in some ways since this study was issued, the detailed engineering concepts and voluminous references that he examines in connection with each of these engineering modules represent a valuable starting point for work in this area.

In 1994, the British Standards Institution (BSI) issued a draft "British Standard Code of Practice for the Application of Fire Safety Engineering Principles to Fire Safety in Buildings" (94/340340 DC). This draft code was met with some negative comment because the document did not go further than supply a collection of formulas. Recently, BSI published a "Draft for Development, DD 240,

Parts 1 and 2, "Fire Safety Engineering in Buildings." Part 1 is a guide to the application of fire safety engineering principles; Part 2 gives guidance on the limits of applicability and confidence limits for the equations in Part 1. The original intention with the first publication in 1994 was to prepare a British Standard on Fire Safety Engineering. However, after considering the comments received on the draft code of practice, particularly those concerning the current state of knowledge on the use of fire safety engineering, BSI decided that it should first be published as a Draft for Development before it could be given the status of a British Standard. The intent is to apply this document on a provisional basis, so that information and experience on its practical applications may be obtained.

5.2.4 International Efforts for Code Validation

Currently, a working group under CIB W 14 (International Council for Building Research and Development) has undertaken an effort to validate fire model predictions through a round-robin series of blind fire model predictions (Jones, 1996) which was initiated about March 1995. The international community has chosen a series of nine scenarios, of generally increasing complexity, on which to evaluate the strong and weak points of some 21 existing computer fire models and increase confidence in the use of fire model predictions. The scenarios are

*   single plume under a hood

*   single room with a door opening

*   single room with a door opening into a corridor

*   floor in a hotel or in a health care facility, or both

*   atrium and a room opening into an atrium
*   shopping mall

*   staircase in a multi-floor building

*   very large room

- underground space, room ventilated only from above

Zone models that have been suggested for this (multi year) round-robin evaluation include CFAST, FIRST (an updated HARVARD code), the Japanese BRI-2, the French FLAMME-S2, WPI (another updated version of the HARVARD code), and 11 others. Eight CFD models are entered, including JASMINE (British Fire Research Station), VESTA (French), and KAMELEON (Sandia National Laboratory and SINTEF (NBL Norway)). Only the first two scenarios, of a single plume under an exhaust hood and a single room with a door opening, have been considered so far.

About 1982–1985, a series of experimental fires was carried out in Germany using a surplus nuclear power plant containment. There have been several attempts to model the results, but because of the complex geometry of the compartments and ventilation factors, these attempts have been difficult. However, these containment scenarios are planned to be the basis of a realistic new international program by a subcommittee of the International Standards Organization. The plans are for the modelers each to first use their fire model to try to predict what should have happened. Then the modelers will try to use each other's models. Finally they will be given the measured results of the experimental fires, and asked to find out what modifications to their models would be necessary to obtain accurate results.

The Society of Fire Protection Engineers has established a task group to address computer fire model evaluation. The goal of the task group is to evaluate computer models, intended for use in fire safety engineering, on their applicability, use, and limitations within the evaluation and design processes. To minimize duplication of effort, the task group is using various ASTM guidance documents, i.e., ASTM E 1355, "Standard Guide for Evaluating the Predictive Capability of Fire Models"; ASTM E 1472, "Standard Guide for Documenting Computer Software for Fire Models"; and ASTM E 1591, "Standard Guide for

Data for Fire Models," for the evaluation effort (Meacham, 1996).

A collaborative international program between EPRI in the U.S. and EdF in France (Mowrer and Gautier, 1997) was aimed at comparing the CFAST, COMPBRN III, FIVE, and MAGIC codes discussed earlier with data from the FM/SNL (NUREG/CR-4681 and NUREG/CR-5384) and UL/SNL (NUREG/CR-3192) series as well as a NBS 3-room series (Peacock, 1988).

As pointed out above, real experimental data are needed to validate computer programs and the techniques for using them. A large amount of data exists, but the data are scattered through the literature and are difficult to assess unless the user happens to remember what kinds of tests were carried out for a given program, and knows the name of the author or agency. In addition, new tests by a number of agencies continually create new data.

About 1990, a first attempt at preparing a database that would include some kinds of large fire tests was initiated at the British Fire Research Station, with international participation. On the basis of what was learned, a second-generation framework (FDMS 2.0) now exists for recording and assessing critically evaluated experimental fire data (Portier, 1994). It will be accessible by anyone from the Internet. Data will include both the results of large-scale tests and data obtained with bench scale and laboratory apparatus. It is planned (Portier, 1996) that NIST will implement this comprehensive fire database management system. It will be available in a format (FIREDATA) that can be readily inserted into fire programs for validation of the programs, as well as for a range of other uses.

5.2.5 Features of Some Fire Computer Codes

Table 5.1 lists capabilities of some current computer fire codes described above. Except for field models, however, none does a really adequate job of calculating the impact of a fire on heating and then igniting such targets as cable trays, and probably no code accurately predicts the chilling of the upper layer gas by the large amounts of heat transfer surface and thermal capacity of cable trays in that layer.

The following is a short description of the meaning of some of the column headings.

"Wall Heat Xfer" refers to whether the heat lost to the wall is calculated in the program. Some programs only use an empirical estimate of the heat remaining in the gas, thus greatly reducing the amount of calculation per time step. As mentioned above, most programs that do the calculation consider only the walls and ceiling as heat loss surfaces, ignoring the effect of other structures in the hot gas layer, such as cable trays.

"Lower Level Gas Temp?" refers to whether there is provision for upper layer gas to mix with or radiate to heat the lower layer of gas.

In all cases, except for COMPBRN III, the "Fire" is entered as input. This column refers to whether it has a constant heat generation rate, or can vary with time, and whether there can be more than one fire in a compartment.

"Gas Concentrations?" must be specified as emissions from the fire vs. time if the program is expected to keep track of them from compartment to compartment. Most of the programs listed on Table 5.1 will perform that task. "Oxygen Depletion" refers to whether the program will shut off or otherwise diminish the fire if the oxygen concentration gets too low for combustion to take place. However, the data for modeling the effect oxygen depletion has on the burning rate are generally not available.

It is assumed that any multi-room model has connections (doors) horizontally on the same level between rooms, and doors or windows from rooms to the outside. Only some of the models can cause gas to flow vertically from a room to one above or below it. This is indicated in "Vertical Connections?" Likewise, any multi-room model (except the smoke flow models) has buoyant flow of gas from one room to another. But only some of them can add forced flow from the heating, ventilation, and air conditioning (HVAC) system ("HVAC Fans and Ducts").

"Detectors?" refers to whether the model will calculate the time at which a thermal detector (including the actuating strut in a sprinkler) or a smoke detector will actuate. The "Sprinklers?" column refers to whether the model will throttle the fire as the sprinkler water impinges on it after the sprinkler strut actuates.

Many other aspects of each model must be taken into account when selecting one for a particular case. With the current models, the general caution is that the strengths and weaknesses of the model must be known to the modeler.

5.3 CONCLUSION

Review of developments in the nuclear industry in France revealed a significant effort and program underway to develop and use fire computer models for determining the risk from fire events. The French program includes research work for fire code development and validation with tests, and application of the developed fire computer code in the fire PRA studies initiated in 1993. The goal of the French program is to advance the state of the art of fire models for nuclear plant applications beyond the current state worldwide, including the U.S.

The review of developments in the U.S. and foreign building industries indicated a notable move toward the use of performance-based methods, and, to a limited extent, risk analysis to replace current prescriptive requirements. Recognizing the benefits of performance-based methods, several countries (New Zealand, Australia, Canada, and U.K.) have modified their building fire laws and regulations to make this transition to performance-based regulation. Australia and Canada are pursuing the use of risk analysis in conjunction with performance-based methods for building fire protection design. More recently, the National Fire Protection Association in the U.S. has also initiated development of performance-based standards. Several insights can be gained from the experience of the building industries for developing performance-based regulations for fire protection of nuclear power plants.

Since the early 1980s, notable developments have been made for fire safety engineering analysis for building safety using fire models, particularly in the U.S., U.K., and Japan. A number of computer codes have been developed and are currently being used for analyzing building fire protection. Recently, an international collaborative effort involving several countries has been initiated to validate fire computer codes being used in the different countries. Several international conferences are now held annually to present and share results and experiences. Other than efforts in France, a similar level of international activity for developing the capability for performance-based analysis for nuclear power plant fire protection is not evident. One collaborative effort between U.S. and French utilities to compare fire computer codes is noted.

Table 5.1 Features of Several Fire Computer Codes

Program	Type	No. Rooms	Wall Heat Xfer	Lower Level Gas Temp?	Heat Targets?	Fire	Gas Conc?	O₂ Depletion	Vertical Connection?	HVAC Fan & Ducts	Detectors?	Sprinklers?	Remarks
CFAST	Zone	15	Yes	Yes	No	Specified -multiple	Yes	Yes	Yes	Yes	Yes	Yes	Fewer rooms if PC
FASTLITE	Zone	3	Yes	Yes	No	Specified	Yes	Yes	Yes	Yes	Yes	Yes	Easy input & run for PC
COMP-BRN III	Zone	1	Yes	No	Yes	Growth calc.	No	Yes	No	No	Yes	No	Input distributions for Monte-Carlo calculations
FIVE	Provides initial screen, leads to use of PRAs, look up tables												Gathers info & keeps records, no computer necessary
FLAMME	Zone	Multi	Yes	Real	Yes	Specified multiple	Yes	Yes	No	Yes	No	No	French, IPSN
MAGIC	Zone	Multi	Yes	No	Yes	Specified multiple	Yes	Yes	?	?	?	?	French, EdF
FLOW-3D	CFD	Few	Yes	Real	Yes	Specified	Yes	Yes	Yes	Yes	Yes	?	Depends on user, significant computing time, & acceptable granularity
LES	CFD	Few	Yes	Real	Yes	Specified	Yes	Yes	Yes	Yes	Yes	Yes	
FPETOOL	Zone	2-1/2	No	No	No	Specified	Yes	Yes	No	No	Yes	No	Easy input for PC, has "TOOLS"
ASCOS (Smoke Flow)	Network Flow	Multi	No	NA	No	NA	No	NA	Yes	No	NA	NA	ASHRAE Documentation
CONTAM (Smoke Flow)	Network Flow	Multi	No	NA	No	NA	Yes	NA	Yes	No	NA	NA	Superior numerics, front end and graphics

6 APPLICATIONS OF RISK-INFORMED, PERFORMANCE-BASED METHODS

In a broad sense, risk-informed and performance-based methods can be thought of as a means of providing an alternative option for implementing regulations that is more efficient in terms of expenditure of resources, while at the same time focusing proper attention on the risk-significant aspects of the regulation. This means may potentially be achieved by an increase in risk-informed discrimination offered by the methodology presented earlier in this report.

The two main objectives of risk-informed and performance-based approaches are

(1) to provide flexibility by emphasizing the safety objective rather than the means for achieving the objective

(2) allocating resources to the most risk-significant areas and minimizing resource allocation to areas in which safety benefit is minimal

In Chapter 3, a comprehensive review of 1351 exemptions and deviations to current fire protection requirements and guidance documents was presented to determine the experience with current prescriptive requirements. The areas that may be amenable to risk-informed, performance-based methods to improve the regulatory process were also determined

This chapter presents several trial applications, or case studies, to examine the potential of risk-informed, performance-based methods (discussed in Chapters 4 and 5) to provide new or improved insights for fire protection analyses, and a more systematic process to judge the acceptability of alternative approaches in some of the areas identified in the exemption review. The applications are presented to examine benefits, and illustrate the manner of potential applications. The material in this chapter may be used as information toward the formulation of regulatory guidance for the applications presented below, but it will be necessary to further define the specific framework of the applications, including

identifying the bounds of validity for the methods for specific cases.

6.1 CATEGORIZATION OF METHODS AND APPLICATION AREAS

The following is a categorization of the methods and application areas. The experience with current requirements is presented in each method category with summaries of the requirements and the exemptions to those requirements, including the technical issues considered by the staff when granting those exemptions. As indicated in Chapter 3, which presents a detailed review of experience with current requirements, the justifications provided for the exemptions, and bases for granting them, were mostly qualitative analyses and engineering judgment. The summary of the experience is followed by a description of risk-informed, performance-based methods that are now available. Detailed examples that apply the methods in several areas of current requirements are presented in Section 6.2, "Applications."

6.1.1 Performance-Based Methods

The first general category of methods is those that would support performance-based approaches, but are not necessarily risk-informed, i.e., these methods will support implementation of less-prescriptive safety objectives, but do not directly analyze or utilize risk information.

Application Areas

1. "Engineering Tools" for Evaluating Fire Dynamics and Use of Fire Computer Codes Based on Zone Models

Section III.G.2.a of Appendix R requires the use of 3-hour-rated barriers to separate redundant trains. Structural steel forming a part of, or supporting, these barriers must also be protected to provide a fire resistance equivalent to that of the barrier. Section III.G.2.b requires separation of cables, equipment, and associated circuits of redundant trains to be accomplished by a horizontal distance of more than 20 feet with no intervening combustibles. In addition, fire detectors and an automatic suppression system must be installed in the area. Section III.G.2.c provides another compliance method to protect safe-shutdown capability by enclosing one of the redundant shutdown trains in a 1-hour-rated fire barrier, and providing fire detection and automatic suppression in that area. Section III.F requires automatic fire detection in areas containing safety-related systems.

A total 624 of exemptions have been given for unrated components (watertight doors and steel hatches), barriers with unprotected openings, partial barriers or less-than-3-hour-rated barriers (e.g., dampers and doors), intervening combustibles within the 20-ft separation, no automatic suppression with low fire loading and high compartment ceilings, no automatic fire detection in areas containing safety-related equipment, and no fixed fire suppression for areas (e.g., control room).

The staff considered that exemptions were acceptable for configurations with low fire loading (including transient combustibles), if the fire severity (measured in minutes) is much less (by a factor of 2 or more) than the installed barrier. Availability of fire detection, auto-suppression, barrier location, and room geometry were also considered for determining the adequacy of barriers. Manual actions for replacing, restoring, or regaining control of a system being protected from a fire with barriers was credited when determining adequacy of the barriers if detailed fire procedures for the actions were available and if the likelihood was high for successful implementation of these actions. Fixed fire suppression was required unless the combustible loading was low, fire detection was available, and the area was continuously manned or sufficient time was available for manual suppression considering a propagation rate of the fire. Except for two exemptions, most of the justifications provided by licensees with the exemption request, and bases for granting them by the staff were based on qualitative analysis and engineering judgment.

"Engineering tools" based on the principles of thermodynamics, fluid mechanics, heat transfer and combustion have now become more available and can be useful for analyzing unwanted fire growth and spread (fire dynamics). The use of these methods will require an evaluation of their validity for specific applications. These analyses can be mostly conducted by hand without a computer program, or sometimes with simple computer routines of fire correlations. "Engineering tools" are available for calculating an equivalent fire severity, adiabatic flame temperature of the fuel in comparison to the damage temperature of the target, fire spread rate, pre-flashover upper layer gas temperature, vent flows, heat release rate needed for flashover, ventilation limited burning, and post-flashover upper layer gas temperature.

These tools can be used to evaluate the adequacy of deviations from prescriptive requirements for configurations with low fire loading, or to establish the basis for fire barrier ratings, safe separation distance, and need for fire detectors and suppression systems in protecting one train for safe shutdown. Since these tools mostly employ bounding calculations, (it will be necessary to examine this for each specific application), results will be conservative but can provide useful information to indicate areas where fire protection features have been grossly overemphasized (or underemphasized).

In cases in which hand calculations are determined to be bounding and conservative but cannot be used to provide useful results, fire computer codes (e.g., FPETOOL, CFAST, COMPBRN) can be used if more detailed calculations are necessary to support a more realistic assessment of the fire hazard and predict fire protection system response. These computer codes are based on plume correlations, ceiling jet phenomena, and hot and cold layer development and can predict the temperature of targets exposed to fires, detector and suppression system actuations, and smoke level and transport during fires. Complex computer codes are used in other areas of NRC regulations, e.g. for simulating thermal-hydraulic, neutronic, and severe-accident transients

2. Reliability Methods

Section III.J requires emergency lighting units with a minimum 8-hour battery-powered supply for all areas needed for the operation of safe-shutdown equipment. These supplies are tested for performance to verify the capability to supply 8 hours of battery power. Several National Fire Protection Association (NFPA) standards prescribe methods and intervals for automatic fire and smoke detector, and suppression system (including fire extinguishers, hoses, and pumps) maintenance and surveillance.

Using the 50.59 change processes, a few plants have modified their fire detector and suppression system maintenance and surveillance intervals, and emergency lighting testing program. Exemptions are not required for these changes.

The main issue considered by licensees in making the modifications to their emergency lighting program was whether an adequate level and duration of illumination with high reliability (including consideration of availability and operability) was provided. Maintenance and surveillance schemes for protection fire detectors and alarms have been modified and set at optimal intervals based on consideration of reliability and performance. Results of the surveillance program are then analyzed to demonstrate that an adequate level of reliability and performance has been achieved, and if necessary, the established maintenance and surveillance intervals are adjusted.

Several reliability-based (based on operating data) methods are available now and are being used in other areas of NRC requirements. For example, NRC requirements in Appendix J of 10 CFR Part 50 allow licensees an option to formulate a performance-based program for containment leakage testing (NUREG-1493). Such approaches can be used to determine an optimal and adequate maintenance and surveillance test interval for fire protection detection and suppression systems. Reliability methods can also be used to demonstrate that testing 8-hour battery-powered supplies for performance at less than full capacity (e.g., 5 hours) indicates a high reliability for their performance at full capacity (8 hours). Currently,

there is no guidance or standard for the use of reliability methods in nuclear power plant fire protection programs.

6.1.2 Risk-Informed, Performance-Based Methods

The second general category of methods is those that would support performance-based and more risk-informed approaches, i.e., methods that will support implementation of less-prescriptive performance criteria, and that analyze or utilize risk information.

1. Use of Risk Insights in a Qualitative Manner

Section III.J requires emergency lighting units with a minimum 8-hour battery-powered supply for all areas needed for the operation of safe-shutdown equipment.

A total of 39 exemptions have been given to allow no lighting in certain plant areas typically inside containment or in the yard; some exemptions applied to indoor areas outside the containment. The key consideration used by the staff for determining if the exemption should be granted was whether emergency lighting was provided for the fire area with sufficient duration so that manual actions that may be required to be performed in the area, based on emergency plans and procedures, can be completed within that time. Although qualitative concepts similar to those in human recovery models used in probabilistic risk assessments (PRAs) were provided as justification for the exemptions, more rigorous PRA models, including modeling of human recovery actions, were not submitted with the exemption requests.

The results of PRAs and other individual plant examination for external events (IPEEE) analyses, including human recovery modeling, and other more limited analysis (e.g., using the FIVE method) are now available and can be used in a qualitative manner to provide risk insights regarding the impact of alternate approaches. An example is the use of fire PRA results, including human recovery modeling, to develop the basis for the plant emergency lighting program in lieu of prescriptive requirements (e.g., 8 hours' duration for all plant areas containing safe-shutdown equipment). Risk-significant accident sequences (e.g., for fire-induced station blackout) can be examined to determine the need for emergency lighting. In some cases, lighting may be required for more than 8 hours.

2. Risk-Graded Approach

Section III.G requires that all structures, systems, and components (SSCs) of one safe-shutdown train be protected from fires by the same measures, regardless of the extent of vulnerability of those SSCs to a fire or impact on plant risk if they are damaged.

A total of 780 exemptions have been given (Section III.G) by the staff for SSCs that have a low vulnerability to fires, or other means for coping with the fires are available so that one safe-shutdown train is protected from the effects of fires commensurate with the risk associated with fire damage to those SSCs.

Fire PRA and other methodologies have inherent in them screening processes that can progressively distinguish between and identify high- and low-risk fire areas. The screening methods employed in fire PRAs, and other methods such as FIVE, can be used toward formulating a risk-graded fire protection program by identifying and focusing on critical fire areas. Categories, or grades, can be established for currently identified fire areas in plants. A higher level of fire protection could then be extended to fire areas that contribute significantly to plant fire risk. This approach would be in contrast to prescriptive requirements that specify that all SSCs of one shutdown train be protected from fires by the same measures, regardless of the extent of vulnerability of those SSCs to a fire or impact on plant risk if they are damaged.

3. Delta-CDF Calculations

Section III.G.1.a requires that the installed fire protection features be capable of limiting fire damage so that one train of systems necessary to achieve and maintain the hot-shutdown condition is free of fire damage. Section III.G.1.b requires

that systems necessary to achieve and maintain cold-shutdown must be repairable within 72 hours. Section III.L contains the requirements for an alternative and dedicated shutdown system and requires the capability to reach cold shutdown (with onsite power only) within 72 hours.

A total of 53 exemptions have been given for approved repairs of hot-shutdown equipment that range from simple low-voltage fuse pulling (to prevent spurious operation) to more complex actions that involve lifting leads and attaching jumpers to permit local operations, and for the use of nonstandard system alignments over a protracted time (more than 72 hours) to reach cold shutdown.

The staff considered that exemptions could be granted if one division was available free of fire damage with allowance for only simple repairs, for which tools, materials, procedures, and staffing were controlled and readily available, and that required a time period for reasonable assurance of success that was much less than the time period in which the component or system being repaired for safe shutdown would be needed. The use of nonstandard systems for achieving cold shutdown over a protracted time was permitted by the staff only if it was demonstrated to have a reasonable chance of success. However, the decisionmaking process only included qualitative analysis and engineering judgment.

PRA operator recovery models and delta-CDF calculations are now available and can be used to supplement the information used to determine the adequacy of alternative approaches. Regulatory guides currently being developed for implementing specific changes to a plant's licensing basis allows the use of delta-CDF as an indicator of the acceptability of implementing specific changes. Fire PRA methods can be used to calculate the change in core-damage frequency (delta-CDF) for alternative approaches to fire protection, including for evaluating the role of operators for recovery actions. These methods are useful for evaluating the extent to which repairs are appropriate to maintain one train of systems to achieve and maintain shutdown conditions, and the use of non-standard systems for shutdown.

The methods can also be used to evaluate and compare alternate means of providing fire protection (by combining separation, fire barriers, and detection and suppression) to safe-shutdown systems.

6.2 APPLICATIONS

The following applications were conducted and are presented to illustrate the benefits of applying the methods that are now available and the subject of this technical review:

A. Performance-Based Analyses

- "Engineering Tools" for Evaluating Fire Dynamics—Bounding Analyses of Combustible Fire Loads

- Reliability Methods

  - Establishing Surveillance Intervals Based on Performance and Reliability

  - Optimizing Test Duration for Appendix R Emergency Lighting

  - Considerations for the Use of Portable Lights for Outdoor Activities

- Fire Computer Codes Based on Zone Models—Analysis of Safe Separation Distance

B. Risk-Informed, Performance-Based Analyses

- Use of Risk Insights in a Qualitative Manner
  - Evaluation of Need for Emergency Lighting

- Event Tree Modeling and Delta-CDF Quantifications

  - Analysis of the 72-Hour Criterion To Reach Cold Shutdown

  - Evaluation of Loss-of-Offsite-Power Assumption for Alternative or Dedicated Shutdown Capability

These applications are presented as examples in the next few sections. Details of the analyses for some of the applications are presented in Appendix D.

6.2.1 Performance-Based Analyses

6.2.1.1 "Engineering Tools" for Evaluating Fire Dynamics—Bounding Analysis of Combustible Fire Loads

In many cases, configurations with low fire loadings (including transient combustibles) can be distinguished from high-risk areas through the use of "engineering tools" that represent fire dynamics in a gross manner. The following is an illustration of how simple tools can sometimes be sufficient to predict the degree of threat from fires by producing credible and useful results. A cable spreading room in a nuclear power plant toured by the authors is used as an example.

The room is about 6.1 m (20 ft) × 6.1 m × 5.2 m (17 ft) high. The upper half of the room is crowded with cable trays, each of which has an array of cables. A fire can only occur with a "transient" fuel, such as spilled cleaning fluid. Assuming a worst-case situation in which the liquid fuel pool is directly below the lowest cable tray, a plume correlation in FPETOOL (a compilation of correlations for fire protection calculations discussed in Chapter 5) can be used to estimate the temperature of the plume at the 3.1-m height of the tray for a series of fire sizes. If it is assumed that the wire insulation will start to degrade at 200 °C, and the fuel would burn long enough for the insulation to reach the plume temperature, the corresponding fire size from the correlation is 400 KW. If the fuel is gasoline (most solvents used for cleaning have a significantly lower burning rate than gasoline, e.g., methyl alcohol burns at 1/4 the rate of gasoline), one can use correlations developed for hydrocarbon pool fires in the SFPE Handbook of Fire Protection Engineering to determine that the pool would be about 1.1 m (3.5 ft) in diameter and the liquid surface would burn at about 4.5 mm/minute (7.5×10^{-5} m/sec) (from Figures 3-11.2 and 3-11.3 in the SFPE Handbook). The volume of the fuel can be determined from the following correlation for the maximum pool diameter (Equation 11, pg 3-203 in the SFPE Handbook):

$$D_m = 2[V^3 g'/y^2]^{1/8}$$

where g' is the effective acceleration due to gravity = 9.8 m/s², y = fuel burning rate (m/s). Solving for V, V = 1.9 x 10⁻⁴ m³ = 0.2 liter.

However, this pool, about 2.5 mm thick, will only burn for about 4 seconds, which is insignificant compared to the time that would be required to heat the lowest cable tray to near the plume temperature. This examines the importance of this fire scenario; others will also need to be evaluated. These calculations can provide useful information toward plant decisions in terms of the degree of fire protection necessary for different configurations and thermal loads.

The tools allow using some information representing the fire dynamics of the problem, and can be used to prevent overemphasis (or underemphasis) that can occur when such considerations are omitted and the hazard from all fire areas regardless of the fire source are equally treated.

Based on this type of analysis, plant procedures need not control transient fuel below a certain volume for which it can be determined that the hazard is negligible. For such purposes, it will be necessary to determine that the correlations used are valid for the specific application, and that results obtained are bounding for the spectrum of fuel spills possible and the hazard from the spill. Currently, a compilation of such tools for applications in nuclear power plants does not exist. Although a broad spectrum of applications has not been explored in this study, it is judged that a sufficient number of applications are possible and an effort to compile these will be useful by providing licensees additional flexibility in maintaining their fire protection programs.

6.2.1.2 Reliability Methods

This section presents the application of reliability methods (feedback of basic performance experience or formal modeling) for determining surveillance and testing schemes for equipment and components in a nuclear power plant fire protection program. The NRC requires that each licensee specify in the plant's technical specifications or fire protection program the surveillance schedules for fire protection equipment and installations in the plant. Specified surveillance intervals similar to those in the relevant deterministic and prescriptive National Fire Protection Association (NFPA) consensus guidelines or standards have been endorsed by the NRC in the past. The Oconee Technical Specifications were examined and the surveillance requirements were compared with the relevant NFPA standards as shown in Table 6.1.

Optimizing surveillance intervals in nuclear power plants on the basis of performance and reliability considerations is an important objective because of the potential for reducing occupational exposure received during the surveillance, especially within the reactor building, where inspections involve donning protective clothing, dosimetry, and decontamination of detectors that are removed for inspection.

The impact of surveillance frequency on the performance (reliability) of standby components has been the subject of many reliability analyses (NUREG/CR-5775). In a performance-based testing approach, surveillance intervals are set based on performance and equipment reliability. If formal reliability methods are used, engineering information is needed regarding the types and the extent of the faults detectable by the surveillance activities (surveillance effectiveness), and the probabilities or the failure rates associated with the occurrence of such faults.

Applications in three areas are presented below: (1) Methods ranging from simple analysis of performance data to using reliability models to optimize test and inspection intervals for fire and gas detectors, and fire valves and extinguishers; (2) A reliability approach for determining the optimal duration for Appendix R emergency lighting tests, and (3) Reliability considerations for the use of portable lights for outdoor activities.

Table 6.1 Comparisons of Fire Protection Equipment Surveillance

Surveillance	Tech Specs	NFPA Code
Fire pumps, 6000 gpm each		NFPA 25
Functional test of pump	Monthly	Weekly
Check proper valve alignment	Monthly	Weekly
Verify flow >3000 gpm	Annually	Annually
Complete system flow test	1–3 years	Annually
Sprinkler and spray system		NFPA 13
Functional test	Annually	Annually (some valves)
Inspect spray area (no obstruction)	Refueling	Monthly
Inspect spray header nozzles	Annually	Annually
Fire hose stations		NFPA 25 and 1962
Visual inspection	Monthly	Monthly
Remove and re-rack hose	Annually	Annually
Check valve	1–3 years	----
Hydrostatic test	1–3 years	1–3 years
Visual inspection (reactor bldg.)	Refueling	Annually
Detectors		NFPA 72
Test operability	Semiannually, some parts quarterly or semiannually	Annually
Carbon dioxide systems		NFPA 12
Check each valve	Monthly	Mfg. recommendation
Check CO_2 tank weight	Semiannually	Semiannually
Verify operation	Refueling	Annually
Flow test (no blockage)	Refueling	Annually

6.2.1.2.1 Establishing Surveillance Intervals on the Basis of Performance and Reliability

The authors visited Catawba, the newest Duke Power Company (DPC) plant, to investigate initiatives being pursued there to optimize and improve the fire protection program. Although there are some case-specific Appendix R requirements, the plant is considered to be a Standard Review Plan plant for the purposes of fire protection regulations. Since the fire protection requirements were moved out of the Technical Specifications into the Final Safety Analysis Report (FSAR), most of the programmatic changes can be implemented with the 50.59 safety evaluation process. However, the operating license requires that the Catawba fire protection program be maintained as stated in the Safety Evaluation Report (SER). Specific fire protection commitments that are cited in the SER or as license conditions can require the license amendment process to implement a change.

Among several other initiatives, DPC examined optimizing the surveillance interval for valves in fire protection systems. At Catawba, approximately 48 hours a month (for both units) are spent confirming fire protection system valve positions. About 400 valve sites are inspected.

The valves are locked and under operations key control. In 3 years, none of these valves have been found in the wrong position. Using the safety evaluation process, DPC has proposed increasing the surveillance interval on the basis of past experience. The inspection interval would change to quarterly, semiannually, and (finally) to annually if the plant maintains a more than 99 percent success rate. A similar approach was pursued for determining surveillance intervals for fire extinguishers and other fire protection components. It is noted that although such initiatives were pursued by DPC, these type of initiatives are not typical in U.S. nuclear power plants because of the lack of guidance and a standard for implementing such performance-based approaches. These performance-based applications can be used as a model for developing regulatory guidance.

The use of reliability engineering models supported by actual failure data for evaluating appropriate test intervals for fire detectors has also been considered in a domestic nuclear power plant. A study reported for Nine Mile Point Nuclear Station Unit 2 (NMP2) is an example of such an activity (Bruce, 1995). Fire and gas detectors in safety-related areas must be tested periodically to detect dormant failures, that is, to check that they will respond if there is an actual demand. Currently, these detectors are sometimes tested as frequently as every 3 months in most plants (e.g., see Table 6.1). Records over a period of five years of fire detector testing were utilized to establish plant-specific fire detector failure rates. Three types of detectors were considered—ionization, heat, and photoelectric detectors. The surveillance records covered 3 years of semi-annual test intervals, followed by 2 years of annual test intervals. An alternative testing methodology was proposed and implemented by the utility based on a 10-percent rotating sampling at the annual test interval, with provisions for expanded sample population upon discovering one or more detector failures. Again, this type of initiative is not typically found in U.S. nuclear power plants because of the lack of guidance and a standard for implementing such performance-based approaches.

A study (Hokstad et al., 1995) published in Reliability Engineering and System Safety by SINTEF (Stiftelsen for Industriell og Teknisk Forskning), uses a detailed reliability model for optimizing the test schemes for fire and gas detectors in a nuclear power plant. The study was performed in three steps. In the first step, the detector failures were classified into random, test-generated, and test-independent faults. The selection for this type of classification was driven by the reliability model developed for this study. In the second step, the effectiveness of various test strategies in detecting the failures in terms of probability of detection was determined. This is an important step because not every failure mode can be detected with one type of test. This is a deviation from standard reliability models, which assume a specific test is perfect (detects all types of failure). Finally, the parameters of the reliability models (including the uncertainties) were estimated through statistical techniques. These parameters then were used in the reliability model for optimization of the test strategy.

The study concluded that the functional test interval extended from quarterly to annually provides better reliability performance at a lesser cost if it is supplemented by daily self-verification and quarterly inspection. The importance of expert judgment in the analysis, which was quite informal in this application, was noted. This study presents one approach for surveillance optimization using reliability performance analyses techniques. Such techniques and evaluations can be applied in a variety of situations in fire protection areas (e.g., suppression surveillance)

An important note on the methods used for establishing surveillance intervals based on performance and reliability is that they do not involve the uncertainties normally associated with fire models and risk assessment, and therefore are subject to less limitations. Performance-based methods have been successfully demonstrated for the testing of containment systems, isolation valves, and penetrations, and there is very little difference between the performance-based analysis methods for such testing programs compared to those that could be used for fire protection systems (e.g., fire valves, pumps, and detectors). Similar benefits for optimizing the testing program by focusing on those components that exhibit poor performance can be derived for performance-based testing programs for fire protection systems.

6.2.1.2.2 Optimizing Test Duration for Appendix R Emergency Lighting

Section J of Appendix R to 10 CFR Part 50 requires that emergency lighting units with an 8-hour battery supply be provided "in all areas needed for the operation of safe shutdown equipment and in the access and egress routes thereto." The intent of this requirement is to allow safe evacuation and the fire fighting activities required to extinguish a fire in an area, and to facilitate operator actions in indoor and outdoor locations if normal and emergency plant lighting are not available after a fire. The prescriptive requirement for an 8-hour duration of lighting was based on conservative engineering judgment and was reasonable given the state of the art for fire assessments and probabilistic risk assessments when Appendix R was instituted. Since that time, licensee experience with the 8-hour battery requirement, both indoors and outdoors, has prompted its reexamination.

Experience with this requirement is summarized as follows: Appendix R emergency lights are tested annually for full 8-hour rating. Many of these lights (about 30 to 50 percent) typically fail during the test after 6 to 8 hours. The dominant failure mechanism is reported to be the depletion of the battery. The fraction of failures and, therefore, the cost for replacement and testing can be significantly reduced if the duration of the test is reduced from 8 to 5 hours.

The following is a quote from NUREG/CP-0129:

> At the Catawba nuclear station, where we have a two-unit plant, we have a total of 50 emergency lights for the fire protection safe shutdown program. We purchased them with 8-hour illumination rating, test them once a year per procedure, and what we find is that about

60 percent of them consistently fail this annual test.[*] The 60 percent that fail normally last 6 hours, 7 hours, or even longer. But because they don't meet the 8-hour endurance test, we have to declare them inoperable, and do a prompt repair. We calculate that we spend about 30 work days a year repairing these lights....

The following analyses examines the benefit of using reliability modeling to investigate the impact of decreasing the emergency lighting test duration from 8 to 5 hours on the probability of failure of emergency lights when demanded.

A variety of battery-operated emergency lighting units are available for use in nuclear power plants. Depending on the battery type used in these units and the quality of the charger, they typically last from 10 to 20 years. Certain types of the batteries, such as lead-calcium batteries, will have a much shorter lifetime if they are frequently discharged. The batteries usually are designed with about a 25-percent safety margin; that is, an 8-hour-rated battery, when equalized and new, may have a discharge time of up to 10 hours. However, when the battery has experienced a full discharge, the rated capacity will drop proportionally to the number of discharges for 8 hours or some other interval. The potential for using risk information to determine emergency lighting needs for important event scenarios is presented in Section 6.2.2.1.

The rated durations of all types of batteries are a strong function of temperature (Institute of Electrical and Electronics Engineers, IEEE 446). There is a vast amount of information on the effect of severe temperature on battery rating (American National Standards Institute (ANSI)/IEEE 450-1987 and ANSI/IEEE 485-1983). Table 6.2, reproduced from ANSI/IEEE 485-1983, presents data showing the effect of temperature on battery capacity rating.

On the basis of the 40-percent failure reported for Catawba and the preceding discussion, the

---

[*] Catawba has recently reexamined this failure rate. On the basis of newer information, Catawba now estimates an ~40-percent failure rate.

Table 6.2 Effect of Temperature on Battery Capacity Rating [*]

Electrolyte Temperature		Rating Factor Multiplier	Electrolyte Temperature		Rating Factor Multiplier
EF	EC		EF	EC	
25	-3.9	1.52	80	26.7	0.98
30	-1.1	1.43	85	29.4	0.96
35	1.7	1.35	90	32.2	0.94
40	4.4	1.30	95	35.0	0.93
45	7.2	1.25	100	37.8	0.91
50	10.0	1.19	105	40.6	0.89
55	12.8	1.15	110	43.3	0.88
60	15.6	1.11	115	46.1	0.87
65	18.3	1.08	120	48.9	0.86
70	21.1	1.04	125	51.7	0.85
77	25.0	1.00			

[*] Source: ANS/IEEE 485-1983. Reproduced by permission of author.
(1) Correction factors were developed from manufacturers' published data.
(2) This table is applicable regardless of the capacity rating factor used and applies to all discharge rates.

probability of battery failure as a function of discharge duration is postulated to be represented by a normal distribution with a mean equal to the rated capacity[*], and a standard deviation of 0.25 multiplied by the rated capacity. As a battery experiences a number of discharges, the rated duration decreases (typically 5 percent per discharge). With this information, the reliability of emergency lights to operate for 8 hours when tested for 5 hours can be estimated. This estimate can be compared with that of the 8-hour endurance test.

The rating of battery pack emergency lights is described below by a normal distribution with a mean r_m and variance F defined:

$$r_m = [r(1-n_d f_d)][f_c(2)] \qquad (6\text{-}1)$$

$$F = (S)(r_m) \qquad (6\text{-}2)$$

where

r = manufacturer's rating plus 10 percent (e.g., 8.8 hours for 8-hour rated)

n_d = number of full discharges

f_d = derating as a result of each full discharge (0.05)

$f_c(2)$ = one over the rating factor multiplier as a function of temperature 2 (from Table 6.2)

S = safety margin (~0.25)

---

[*] For a new battery with no discharge, a safety margin of 10 percent on rated capacity is assumed, e.g., an 8-hour rating can last up to 8.8 hours when the battery is new.

The probability of failure of an emergency light demanded for D hours as a result of battery depletion then can be calculated from the cumulative normal probability; that is,

$$P_B (D) = N (D: r_m, \sigma).$$

In addition to the battery failures, an emergency light may fail intermittently, regardless of its capacity. The probability of failure of an emergency light as a result of causes unrelated to battery depletion may be estimated from:

$$P_L = \frac{1}{2} \lambda T = P_d \qquad (6\text{-}3)$$

where

λ = random failure rate per year (6.8E-2)[*]

T = test interval (1 year)

P_d = equivalent demand probability (3.4E-2)[**]

The preceding equations can be used for calculating the probability of an emergency lighting failure (P); that is, $P = P_L + P_B (D)$.

In Equation 6-1, n_d stands for the number of full discharges. If the endurance test is performed for a duration less than 0.75 rated value, it is not considered as a full discharge based on battery discharge depth versus life characteristics. When battery-operated emergency lights are installed, they are considered to be new ($n_d = 0$). As the lights are tested annually, some would fail and would subsequently be replaced. After several years, the population of the emergency lights in a given fire area will have different ages (i.e., n_d in Equation 6-1 and τ in Equation 6-4 will depend on the last replacement). A detailed reliability model was developed to estimate the fraction of the lights with different age as a function of time from installation. This reliability model accounts for the probability of an emergency light failing at a given age and being replaced. The model exhaustively calculates all possible combinations.

Figure 6.1 shows the failure probability (or fraction of emergency lights failing during a test), calculated using the above equations, as a function of years after installation for the following cases: 8-hour rated, 8-hour tested (8-R, 8-T) and 8-hour rated, 5-hour tested (8-R, 5-T) at average temperatures of 77 °F (298 K) and 50 °F (283 K). The fraction of lights expected to fail during an 8-hour test at 77 °F (298 K) is about 40 percent, which is consistent with Catawba's experience discussed above. The fraction of lights expected to fail during a 5-hour test at 77 °F is about 15 percent on the basis of the model prediction.

As discussed in Chapter 7, for a unit with about 50 battery-operated emergency lights, 5-hour testing rather than 8-hour testing will result in saving about 12 replacements a year. Figure 6.2 shows the probability of failure upon demand for the following cases: 8-hour rated, 8-hour tested, and 8-hour rated, 5-hour tested for an 8-hour and 6-hour demand at a temperature of 77 °F. These curves show that the reliability performance of the two alternatives are comparable or equivalent (maximum difference in reliability is less than 10 percent).

Finally, Figure 6.3 shows a comparison similar to that in Figure 6.2, but at an average temperature of 50 °F (283 K). Here, the test duration becomes important for a 6-hour demand, and the temperature of the environment becomes an important factor.

A formal uncertainty evaluation was conducted for the preceding analyses to illustrate the availability of techniques to assess the uncertainties in such reliability methods. This evaluation is presented in Appendix D. Hypothetical distributions for the basic parameters were used in this evaluation due to the lack of resources to collect data from the plant and manufacturer; however, such assumptions do not affect the illustration of the techniques which is the purpose of this report.

The above analyses illustrates the type of reliability techniques that may be employed, and the data necessary for providing additional insights when considering modifications to test

[*]From MIL-HDBK-217E (U.S. Dept. of Defense, 1990).

[**]From Bento et al., 1985.

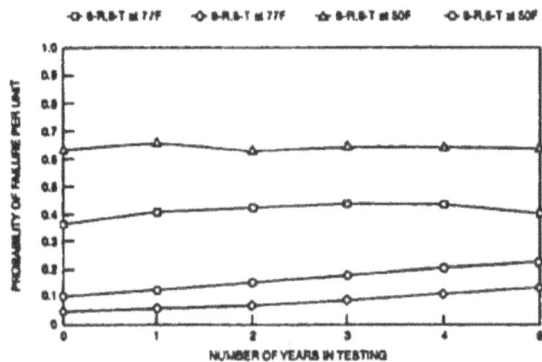

Figure 6.1
Annual Test Failure Probability for Battery-Operated Lights

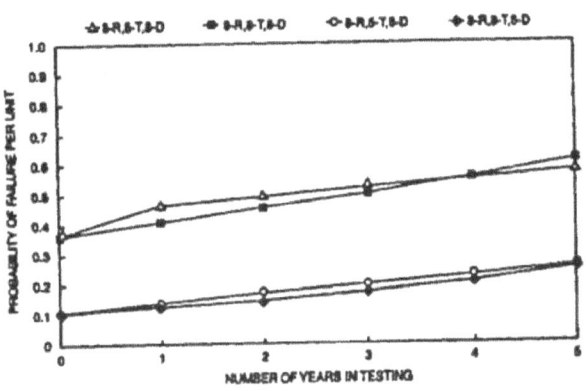

Figure 6.2
Demand Failure Probability for Battery-Operated Emergency Lights (77 °F)

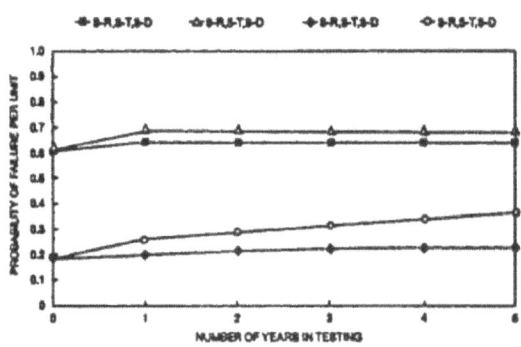

Figure 6.3
Demand Failure Probability for Battery-Operated Emergency Lights (50 °F)

schemes based on performance, such as that for emergency lights. Due to the limited scope and objective of the analysis presented, some assumptions regarding the distribution of the probability of battery failure as a function of discharge duration and the rated value below which the test may not be considered a full discharge were made. Although this analyses estimates that decreasing the test duration from 8 to 5 hours has about a 10 percent impact on reliability and a sensitivity of this reliability impact to temperature, this analysis and result should be considered an illustration. In order to determine the real impact and extent of sensitivities, it will be necessary to collect plant data to test and verify the assumptions made in this analysis.

6.2.1.2.3 Considerations for the Use of Portable Lights for Outdoor Activities

This section does not present an analysis but highlights some considerations for developing a reliability model for the use of portable lights for outdoor areas. Section 6.2.2.1 provides methods for determining the need for outdoor lights. Appendix R requires outdoor emergency lighting to facilitate human actions that are required for safe shutdown. A large number of outdoor lights may be required to get the proper level of illumination necessary for certain actions. The reliability and survivability of outdoor lights, especially in cold winter weather, are questionable. Experience indicates that portable lights, maintained indoors, are a more reliable option than outdoor, fixed, battery-pack lights. Furthermore, some of the human actions that may require the operator to go outdoors may not start until 5 hours after the fire damage has occurred and, depending on the scenario, may last beyond 8 hours. The use of portable lights on an as-needed basis will prolong the availability of emergency lights.

Relief from this requirement has been requested by utilities through submittal of exemption requests. These exemption requests were briefly discussed in Chapter 3 of this report. For example, the Trojan Nuclear Power Plant submitted an exemption request to use portable, emergency battery-lighting units as an alternative to permanent emergency battery-lighting units in selected outdoor locations. Severe winter weather was given as one justification for not using permanently installed battery-operated emergency lights in outdoor areas.

A comprehensive review of available reliability databases indicates similar reliability for portable lights and fixed lights, as long as they are maintained indoors, auto-charged, and under strict administrative control. The selection of one over the other is not based on reliability, but mainly on the type of the task and activity to be performed. The potential need for additional personnel for holding and directing the light beam while a task is being performed is a consideration in determining the effectiveness of portable lights.

6.2.1.3 Fire Models and Computer Codes Based on Zone Models—Analysis of Safe Separation Distance

NRC fire protection regulations require that one train of systems necessary to achieve and maintain hot-shutdown conditions be free of fire damage. The regulation provides three options for meeting this requirement, including one that allows for separation of cables, equipment, and associated non-safety circuits of redundant safe-shutdown trains by a horizontal distance of more than 6.1 m (20 ft) with no intervening combustible materials or fire hazards. In addition, fire detectors and an automatic suppression system must be installed. Analyses were conducted to determine results from the following three fire models for developing insights regarding the 20-ft safe-separation requirement: (1) FIVE—a compilation of fire correlations in worksheets for use in screening fire areas; (2) COMPBRN IIIe—a fire computer code developed for fast computations for use in fire PRAs; and (3) CFAST—a fire computer code developed mainly for use in modeling fires in buildings. These methods were described earlier in Chapter 4 and Chapter 5. The following is a summary of the study. Details of the analyses are included in Appendix D.

A representative PWR emergency switchgear room (ESGR) was used for the study. The room is 15.2 m (50 ft) × 9.1 m (30 ft) × 4.6 m (15 ft)

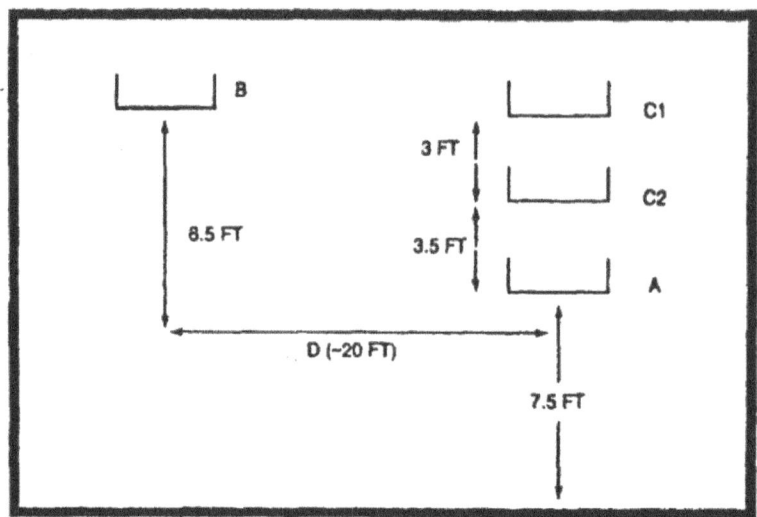

Figure 6.4
Illustration of Critical Cable Locations in the Representative Emergency Switchgear Room
(Configuration 1)

high. The room contains the power and instrumentation cables for the pumps and valves associated with motor-driven auxiliary feedwater trains, all three high-pressure injection trains, and both low-pressure injection trains. A simplified elevation of the ESGR room, illustrating critical cable locations, is shown in Figure 6.4. The power and instrumentation cables associated with safe-shutdown equipment are arranged in separate divisions and are separated horizontally by a distance, D. The value of D is varied in this evaluation. The analysis was conducted for different elevations of Tray B so that it was either in the ceiling jet sublayer or in the hot gas layer for different cases.

The postulated ignition source is either a self-ignited cable (as a result of a fault) or cable ignition as a result of a transient fire. Cable Tray A is considered to be the source. Although, most rooms will be isolated by the automatic closing of fire dampers and the shutdown of the ventilation system, an opening 2 m (6.5 ft) high × 0.2 m (0.7 ft) wide was assumed to prevent pressure buildup in the room and facilitate the use of the COMPBRN and CFAST codes.

The ESGR contains smoke detectors and a manually actuated Halon system. Assuming a performance criterion of 1 hour as the duration in which redundant trains should not be damaged, and considering the fire initiating frequency and suppression (including fire brigade) probability, it can be estimated that the resulting core-damage frequency (CDF) for this scenario will be 1.2E-5 per reactor-year. This core damage frequency and the derivative, 1 hour, during which redundant trains should not be damaged, is used as a criterion to determine the adequacy of the safe separation distance.

The FIVE method predicts that an effective fire source intensity of about 6.5 MW is required to damage cables that are separated by 20 ft, and 3.5 MW if separated by 10 ft, for cables that are in the ceiling jet layer (see Table 6.3). The FIVE screening method does not differentiate between the various separation distances in the hot gas layer and only conservatively estimates, based an adiabatic heating of the gas, the total energy release needed to raise the average hot gas layer temperature to the threshold damage temperature. In the present case, the total energy needed is about 286 MJ, which is much less than 3150 MJ

Table 6.3 Summary Results From FIVE Analyses

Effective Fire Intensity kW	Ceiling Jet Temperature K	Target Damage Temperature K	Separation Distance ft
3500	526	643	20
6500	643	643	20
7000	660	643	20
3500	660	643	10
6500	843	643	10
7000	871	643	10

corresponding to the energy released from a 3.5 MW fire during a 15-minute period. Therefore, none of the cases pass the screening criteria if the target is the hot layer.

The COMPBRN analyses predict (see Table 6.4) that the effective fire intensity, capable of damaging redundant cables separated by 6.1 m (20 ft), is about 4 MW for the representative configuration, and that damage occurs in about 12 minutes. The COMPBRN code also predicts that a cluster of two cable trays in one side of the room (Case 5 listed in Table 6.4) will result in a peak burning rate of about 1.8 MW, which is not sufficient to damage cable trays separated by 20 ft. The heat release rate predicted by COMPBRN for Case 2 is given in Figure 6.5.

A modified version of the CFAST code, which accounts for radiation heat transfer to a target, was utilized for this evaluation. The CFAST code requires input of the heat-release rate for the fire source. Values of 1 MW, 2 MW, and 3 MW with a linear growth taking 1, 2, and 3 minutes, respectively, for the heat-release rate were used for three cases. The hot layer temperature, the radiative and convective heat transfer calculated by CFAST, was used in a transient conduction model for a thin slab to estimate the target surface temperature. Figures 6.6, 6.7, and 6.8 show the hot layer and cable surface temperatures for a 1-, 2-, and 3-MW fire as a function of time. Considering the critical damage temperature of 643 K and the extrapolation of the result shown in Figures 6.6, 6.7, and 6.8, a fire of more than 3 MW is required to damage the target cables at a

20-ft separation in less than 1 hour, and a fire less than 2 MW will not damage redundant cables separated by less than 6.1 m (20 ft).

In order to understand the reason for the difference in the predictions of the CFAST and COMPBRN codes, the availability of oxygen to support the burning rates predicted by COMPBRN (see Figure 6.5) was examined. The CFAST code is capable of calculating the concentration of various species of air and combustible products in the hot layer region, whereas COMPBRN does not have a similar capability. Using burning rates predicted by COMPBRN, CFAST predicts that, at about 5 minutes, the hot gas layer descends to the level of the lowest burning tray and the concentration of oxygen in the hot layer is below 10 percent (ordinary air is 21 percent). Therefore, the heat release rate will not increase after 5 minutes because of oxygen depletion and the fire would eventually be extinguished when insufficient oxygen is available to support combustion. Accordingly, the peak heat-release rate for this specific case will be below 2 MW and the heat-release rate predicted by COMPBRN after 5 minutes is overly conservative.

Figure 6.9 shows a comparison of the results from the CFAST and COMPBRN codes for Case 2 (see Table 6.4 for case conditions). In this case, the heat-release rate due to fire predicted by COMPBRN (Figure 6.5) is provided as input to the CFAST code for the comparison analysis.

Table 6.4 Summary of COMPBRN Results

Tray	Case 1 (Base Case)		Case 2		Case 3		Case 4		Case 5	
	D	I	D	I	D	I	D	I	D	I
I. Damaged (D) and Ignition (I) Time (minutes)										
A (Source)	0	0	0	0	0	0	0	0	0	0
C2	2	2	2	3	2	2	2	2	2	2
C1	4	4	5	5	4	4	4	4	-	-
B (Target)	8	9	9	10	12	No	8	9	No	No
II. Total Heat Release Rate at the Time of Target Damage										
Q, MW	4.8		4.0		8.2		4.7		1.8*	
III. Description of Cases										
Pilot fire size (ft × ft)	4 × 2		2 × 2		4 × 2		4 × 2		4 × 2	
Door	Open		Open		Closed		Open		Open	
Trays above pilot fire	C1 and C2		C1 and C2		C1 and C2		C1 and C2		C2 only	
Target elevation (m)	4.27		4.27		4.27		2.29		4.27	

*Maximum heat-release rate with no damage to target cables.

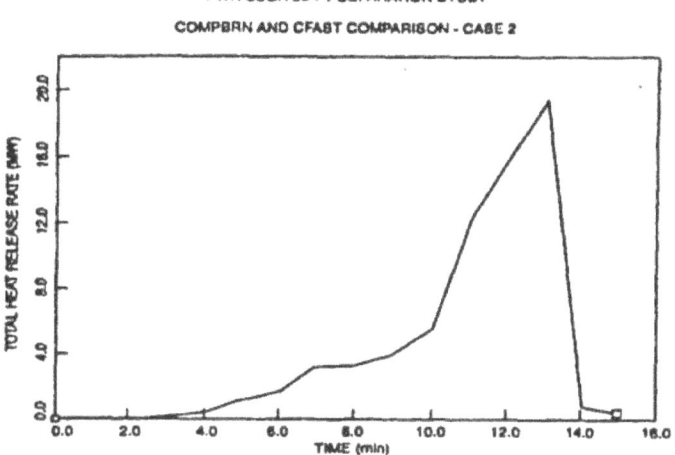

Figure 6.5
COMPBRN-Predicted Heat Release From Burning Cables

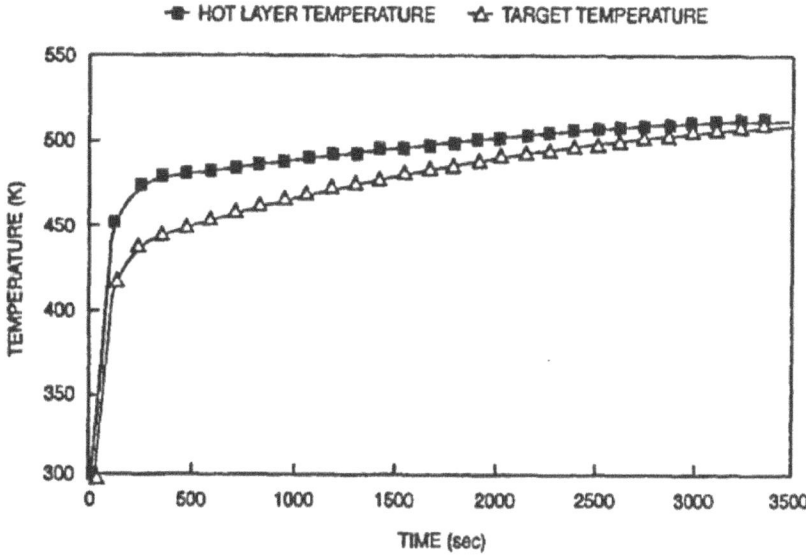

Figure 6.6
1-MW Fire Source Target and Hot Layer Temperature

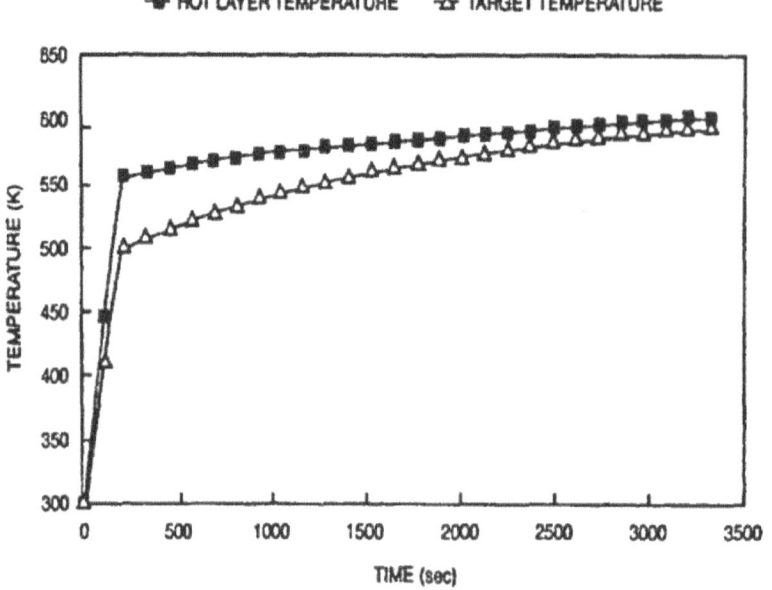

Figure 6.7
2-MW Fire Source Target and Hot Layer Temperature

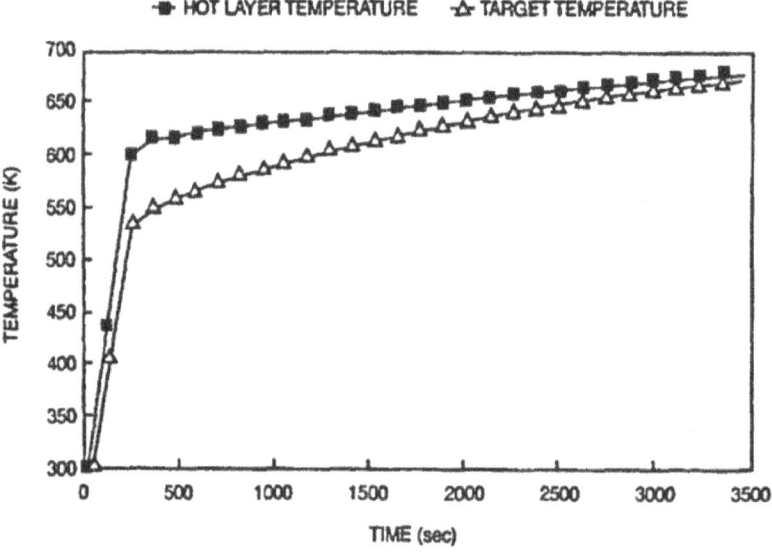

Figure 6.8
3-MW Fire Source Target and Hot Layer Temperature

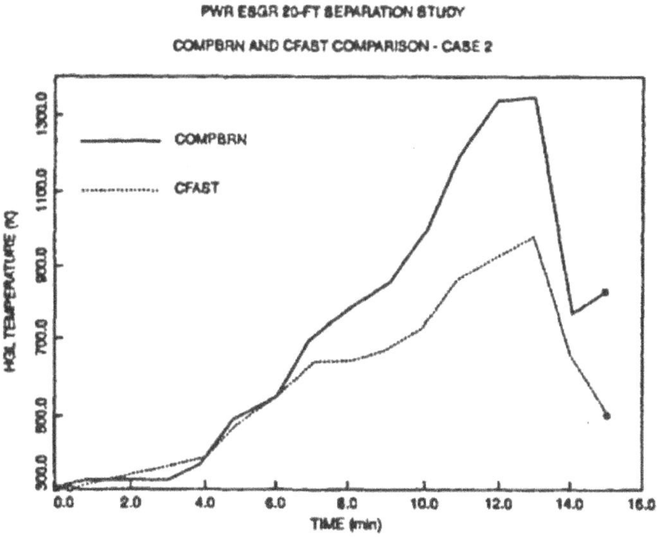

PWR ESGR 20-FT SEPARATION STUDY

COMPBRN AND CFAST COMPARISON - CASE 2

Figure 6.9
Comparison of Hot Gas Layer Temperatures

After the COMPBRN-predicted ignition of Tray C2 at 5 minutes and Tray B (the target tray) at 10 minutes, Figure 6.9 shows that the hot gas layer temperature predicted by COMPBRN is much higher than that predicted by CFAST. This may be due to the conservative assumptions regarding heat losses from the hot layer in the COMPBRN code, however, the reason for this large difference in hot layer temperature was not examined further.

On the basis of the preceding results, it is concluded that if the maximum cluster of source cables results in a heat-release rate less than about 2 MW, then redundant cables will not be damaged, even if they are separated by less than 20 ft (e.g., 15 ft). The dominant factor for all the fire models for predicting damage to cables that are separated by 20 ft is the effective intensity of the fire source, not the total combustible loading in the fire area.

The preceding study illustrates the capability of these fire computer codes to evaluate alternative approaches to the 20-ft separation criteria, although at different levels of resolution. The

FIVE method is formulated for screening purposes, and it does not have sufficient resolution to address the problem in this evaluation if it is assumed the target is in the hot layer. Both COMPBRN and CFAST estimate that a fire of about 1.8 MW or less will not damage redundant cables with 20-ft separation. This corresponds to a maximum cluster of three cable trays.

The preceding paragraph illustrates the type of insights that may be drawn regarding the nature of configurations that are more vulnerable to fire hazards, and the parameters important for such a determination. An analysis of the validity and accuracy of the results is not presented here. Chapter 5 and Appendix C contain some comparisons of the results from computer codes used here, COMPBRN and CFAST, with experimental data. Judgments on the results of the analyses for a specific problem should be made once the validity and accuracy of the models for that application are considered.

6.2.2 Risk-Informed, Performance-Based Analyses

6.2.2.1 Use of Risk Insights in a Qualitative Manner—Evaluating Need for Emergency Lighting

The failure of battery-operated emergency lights when no sources of lighting are available may affect the following plant activities:

- fire fighting activities

- local operator actions

- repair and recovery actions needed to be performed during various scenarios.

The design and operation of the lighting system vary from plant to plant, but the following description provides a general overview.

Indoor Lighting

There is a normal lighting system fed through the onsite distribution system from the offsite power grid. There is also an emergency power source for the lighting system for all fire areas containing safe-shutdown equipment that is fed from emergency diesel generators in case offsite power is lost. The control room typically has additional emergency lighting powered by the station's dc system.

Because a fire could damage normal and emergency lighting for any area of the plant, battery-powered portable lights also are available to facilitate access to and egress from the control room, emergency switchgear rooms, diesel generator rooms, and other areas.

In accordance with the requirements of Appendix R, there is a post-fire emergency lighting system for illuminating all areas needed for operation and for monitoring of safe-shutdown equipment, and to ensure access and egress routes thereto. It consists of self-contained 6-V or 12-V batteries and static charger units located in the area served. This post-fire emergency lighting system will provide sufficient illumination for a minimum of 8 hours to enable an operator to reach the safe-shutdown equipment and carry out the required functions.

Outdoor Lighting

There is not as much redundancy for outdoor lighting as there is for indoor lighting. Usually available are lights fed from offsite power, Appendix R 8-hour lights, portable lanterns, and security lights.

Indoor Emergency Lights

From a safety perspective, emergency lights are used for two types of activities:

Fire fighting. Electric power can be lost to the area that is on fire, thus jeopardizing fire fighting activities. In addition, smoke from the fire can obscure visibility, thus posing further difficulties in performing these activities. The function of the emergency lights is to increase the visibility in both of these circumstances Table 6.5 based on the data in the Oconee PRA (Sugnet et al., 1984) assumes that most fires were extinguished within 1 hour after they were discovered. Therefore, emergency lighting with a duration of more than 1 hour would be sufficient for this aspect of fire safety.

Repair of equipment for safe shutdown. Emergency lights will provide sufficient illumination for a minimum of 8 hours to enable an operator to reach the safe-shutdown equipment and carry out the required functions or repairs. At most plants, the redundant shutdown train is located in a separate area and the lighting will not be affected by the fire (even in case of a loss of offsite power (LOSP) coincident with a fire, the lighting in redundant areas is fed by onsite emergency power). Certain fire scenarios may affect the lighting in both areas; however, this would be limited to a plant-specific vulnerability. In most plants, the most likely scenario for loss of needed emergency lights would be a station blackout (SBO) scenario induced by fire in such areas as the switchgear room, since alternative sources of lighting would not be available.

Table 6.5 Mean Fire-Suppression Time

Mean Fire-Suppression Time (min)	Probability	Cumulative Probability
5	0.10	0.10
15	0.40	0.50
30	0.40	0.90
60	0.10	1.00

Outdoor Emergency Lights

Outdoor lights are normally fed from offsite sources (usually a switchyard), and would not be available in a LOSP transient. Hence, the availability of emergency lights independent of offsite power for outdoor areas, either in the form of portable lanterns or permanently fixed lights, is important.

The following is an example of how insights from risk analyses may be used in a qualitative manner. The LaSalle Unit 2 PRA (NUREG/CR-4832, Vol. 1), directed under the Risk Methods Integration and Evaluation Program (RMIEP), is one of the most comprehensive PRAs conducted to date. In particular, it contains a detailed fire risk assessment (NUREG/CR-4832, Vol. 9), which can be used to develop risk insights about this Appendix R requirement. LaSalle Unit 2 was selected for this example. The LaSalle PRA contained four analyses: internal, fire, flood, and seismic. The total mean core-damage frequency (CDF) from all events reported in the PRA is 1.01E-4 per reactor-year. Table 6.6 shows the relative contributions of accident sequences from the four analyses to the mean integrated CDF. It shows that, together, the internal and fire contributions are 95 percent of the total CDF.

The greatest risk from the failure of the battery-operated emergency lighting (both indoor and outdoor) is incurred during fire-induced LOSP and SBO scenarios, where other sources of lighting are unavailable. Outdoor lights are considered for both the SBO and LOSP scenarios. Indoor lights are considered for the SBO scenario.

Lack of illumination during these scenarios will prevent any recovery or repair or local manual actions.

If credit for all recovery and manual actions (event names starting with RA, OP, and OE) is removed in an extended SBO scenario in the LaSalle PRA (or any other PRA), core damage will occur. However, removing credit for all recovery and manual actions in an extended LOSP scenario in the LaSalle PRA will not result in core damage, unless two additional random failures occur. Therefore, the most stringent requirements for emergency lights will stem from the SBO scenario.

Since SBO (both internal and fire-induced) is the major contributor to the LaSalle CDF, the necessity of emergency lighting is warranted for areas that are affected by a fire or where operator actions will be required to recover from this accident sequence. Various operator and manual actions are required, depending on the scenario of events. In the first 6 hours, when plant dc power is not depleted in an SBO scenario, operator actions will take place in the control room (or remote shutdown panel); potentially in the reactor core isolation cooling room if diesel generator (DG) "2A" fails quickly as a result of DG cooling water failure; in the switchyard to recover offsite power; and in the emergency diesel generator room and the emergency switchgear room, to recover onsite power. After 6 hours (when emergency dc power is depleted), and up to about 27 hours when containment integrity may be challenged, the recovery actions for offsite and onsite power are also questioned in the PRA. The

Table 6.6 Percentage of Total Core-Damage Frequency

Contributor	Percent
Fire	49
Internal	46
Flood	5
Seismic	0

need (in terms of duration) for emergency lighting can be determined from this risk-significant accident sequence.

The requirement for the duration of emergency lighting is a plant- and area-specific issue. Risk insights regarding this issue can be drawn from a plant-specific fire PRA to determine the time available for various manual and recovery actions on a fire-area-specific basis. Generally, the most stringent demand for emergency lighting is imposed by SBO and LOSP scenarios. Emergency lighting may not be needed for manual and recovery actions in those areas for which redundant plant-specific lighting is available and remains unaffected by the fire. An alternative means of emergency lighting using a centralized battery/charger unit may be acceptable for these areas depending on the area-specific features.

The preceding analysis illustrates how information from a fire PRA can be used in a qualitative manner to develop insights on the need and importance of emergency lighting for risk significant and vulnerable accident sequences. A more detailed analysis using plant-specific PRA information can be conducted for examining critical areas for emergency lighting.

6.2.2.2 Event Tree Modeling and Delta-CDF Quantification

6.2.2.2.1 Analyses of the 72-Hour Criterion to Reach Cold Shutdown

In order to limit the amount of repairs to equipment for achieving safe shutdown in the event of a fire, current fire regulations of the U.S. NRC require that a plant have the capability to reach cold shutdown conditions within 72 hours. Experience from the early 1980s in implementing this requirement presented in Chapter 3 indicates that some U.S. plants found it difficult (it would be too costly) to meet this prescriptive requirement and, therefore, requested that they be exempted from this requirement based on qualitative arguments, which indicated that alternatives that included the use of non-standard systems and repairs, and would require more than 72 hours to reach cold shutdown, would provide an equivalent level of safety. These requests for exemptions were based on qualitative analysis and engineering judgment and have been accepted by the NRC (Chapter 6). Since the early 1980s, new methods for fire PRAs have become available and can be used to quantify, through delta-CDF calculations, the impact of using alternative methods for achieving the higher level safety objective. The following illustrates this method. Details of the analyses presented below are provided in Appendix D.

The LaSalle fire PRA analysis for the fire area for the cable shaft room adjacent to the Unit 2, Division 2, essential switchgear room was used for the purpose of this illustration. It was postulated* that the fire area contains equipment associated with both trains of the residual heat removal (RHR) system, and that the fire damage is extensive and it will take more than 72 hours to restore one RHR train. This study adopts the LaSalle PRA assumption that a small fire anywhere in the fire subject area will cause the rapid formation of a hot gas layer that causes all critical cabling to fail. Prescriptive compliance with the 72-hour requirement would necessitate that one RHR train be removed from the fire area, or that it be protected. An alternative approach is postulated to include reestablishing the condenser (power conversion system, PCS) for long-term decay heat removal to allow sufficient time for the repair of one train of RHR shutdown cooling. This approach would take more than 72 hours to reach cold shutdown.

The LaSalle fire PRA used conservative assumptions by excluding credit for operator recovery actions for modeling the subject fire area since it was a non-dominant contributor to the fire-induced CDF. Therefore a more detailed event tree (shown in Figure 6.10) was developed for this example, which included manual actions to recover PCS and RHR. The prescriptive compliance case assumes one RHR train is removed from the fire area or otherwise protected. Therefore, a failure of the containment heat removal (CHR) function requires additional RHR random failures. The estimated unavailability is CHR = 1.1E-1. The alternative case does not protect the RHR system. All containment heat removal is assumed lost due to the fire, and CHR = 1.0. Operator actions to reestablish the condenser and to recover one train of RHR are key actions in this analysis. Detailed plant-specific human reliability analysis would be required to accurately represent important operator actions and potential systems interactions. For illustrative purposes, failure estimates that are more conservative than values

normally used in PRAs were used for these restorations for this study. The four sequences leading to core damage are quantified for both the prescriptive and alternate approaches. The final result is given at the bottom of the figure; it is ΔCDF = 8.0E-7.

The preceding example illustrates the PRA method and the potential for using ΔCDF as a tool toward evaluating the safety equivalence of an alternate approach to a prescriptive requirement. As is the case for this example, alternate approaches can be expected to require reexamination of non-dominant sequences, and use of a finer level of modeling resolution to credit certain operator recovery actions. The purpose of this example was not to only determine a bottom-line ΔCDF (in any case this analysis is not based on a real plant configuration or conditions) but to show that a probabilistic approach provides a systematic framework in which to identify key issues such as operator actions, examine assumptions, sensitivities and uncertainties**.

An important insight derived from the preceding exercise is that most of the risk contribution comes from Scenario 10, which is unaffected by the 72-hour issue. This type of insight provides an indication of the relative importance of issues in the overall plant risk profile.

Since the accident sequences in this application involve key operator actions, the ability of current HRA techniques to model the type of actions involved in these sequences, which may involve several operators over a longer period of time than normally evaluated in current PRAs, must be examined. Sensitivity studies, varying the human error probability (HEP), should be conducted to determine if conservative and bounding values for HEPs are used to validate the insights drawn from the analysis. The dominance of Scenario 10 to the risk contribution, and the uncertainty of continued

---

* It was necessary to assume some changes to the configuration of this fire area in order to allow data from the LaSalle fire PRA to be used for this illustration. Therefore, this analysis does not model the LaSalle plant.

** The results of the uncertainty analysis for this example is presented in Appendix D. It shows that the uncertainty of this analysis is dominated by the uncertainty associated with continued injection after containment failure.

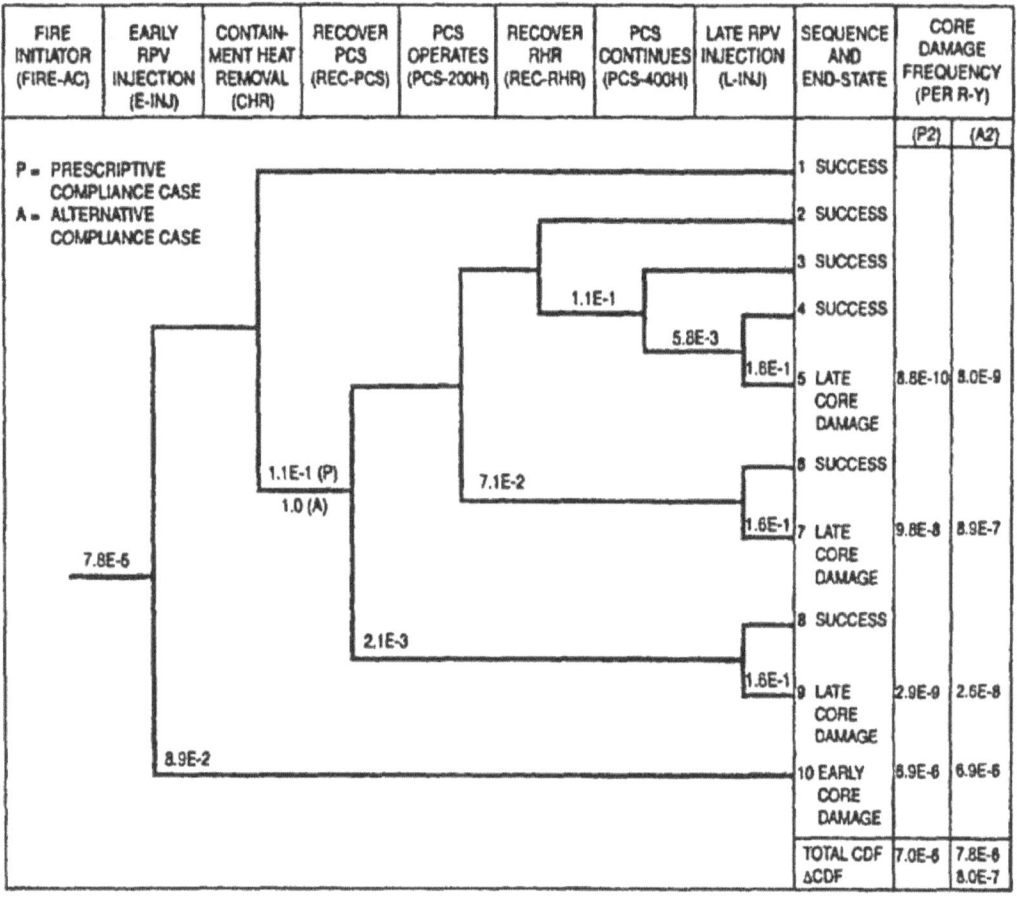

Figure 6.10
72-Hour Case Study—Quantified Event Tree

injection after containment failure to the total uncertainty, provides an indication of the significance of the uncertainty of the HEPs for key operator actions.

6.2.2.2.2 Evaluation of Loss-of-Offsite Power Assumption for Alternative or Dedicated Shutdown Capability

Section III.L of Appendix R requires that an alternative and shutdown capability, if required by criteria established in Section III.G, shall be able to function as intended with a LOSP. The need to postulate LOSP, in conjunction with alternative or dedicated shutdown capability, has been a subject

of discussion since the rule was promulgated. As noted in Chapter 3, several exemptions for alternative shutdown (Section III.L) have been approved by the NRC staff. Noncompliance with the LOSP requirement generally indicates that the plant-specific Appendix R analysis postulates damage to one or more emergency ac power sources. Since the rule requires licensees to postulate a 72-hour LOSP, the critical injection and decay heat removal systems are without power. Within the confines of the Appendix R LOSP requirement, core damage is postulated. This generally requires rerouting or protecting the emergency ac power source to ensure compliance with the rule. Although all operating plants conform to this requirement, continued industry interest in the need to consider LOSP was evident during the workshop on the program for the elimination of requirements marginal to safety (NUREG/CP-0129).

The following analysis illustrates how a fire PRA can be used to provide a systematic process to evaluating the Appendix R, LOSP-assumption requirement. The Limerick auxiliary equipment room, as modeled in the fire PRA (NUS Corporation, 1983), is used for this illustration. The auxiliary equipment room is located one floor above the main control room. This room contains signal-conditioning components housed in steel cabinets, the associated cabling required for the control of all safety-related and balance-of-plant equipment, and the remote shutdown panel. A fire in this area could cause the evacuation of the control room and is expected to require local manual actions for plant recovery.

The Limerick fire PRA examines the consequences of self-ignited cabinet and cable fires and transient combustible fires at various critical locations throughout the auxiliary equipment room. For the purposes of this illustration, however, a single transient combustible fire is postulated at Limerick location b. A fire at this location is predicted to disable redundant systems by simultaneously damaging cables in overhead conduit and the logic circuits in the cabinets. Only train D of the low-pressure coolant injection (LPCI) system (which is served by Division IV ac power) and the capability to depressurize with non-ADS (automatic depressurization system) safety relief valves (SRVs) is assumed available for early accident mitigation. Closure of the main steam isolation valves is expected as a direct result of the fire. All offsite power circuitry is located outside the auxiliary equipment room and remains unaffected by the fire assuming there are no circuit interactions. The support systems for LPCI train D (e.g., emergency service water) are also assumed to be unaffected by the fire. In addition, this analysis, which is aimed at illustrating the method, assumes the Division IV emergency diesel generator cabling is located in the vicinity

of location b', and fire analysis predicts a high probability of damage.

The event tree shown in Figure 6.11 is a quantitative model for the analysis. Two cases are examined. The first assumes prescriptive compliance with the LOSP requirement; that is, the cable tray containing emergency ac train IV cabling and components in the auxiliary equipment room is protected from the effects of a fire at location b. The event tree is quantified using the system success criteria, and the hardware unavailability estimates are based on the Peach Bottom PRA** (NUREG/CR-4550 Vol. 4, Rev. 1, Part 3). The computerized Peach Bottom PRA fault tree models (NUREG/CR-5813) are modified to reflect the prescriptive compliance case configuration. Both offsite and emergency ac power are assumed available. The ac power support system fault tree is further modified for the alternative approach by removing all credit for emergency ac power.

After a major fire in auxiliary equipment room location b, control room evacuation is assumed. The operator will have to reestablish injection by manually depressurizing the RPV and using LPCI train D. The event tree has separated the operator actions to reestablish RPV injection (XHE) from the hardware failures (X) and (V) because these operator actions are interrelated and cannot be easily segregated by system. A failure to reestablish injection will result in early (less than 1 hour) core damage. If early RPV injection (XHE, X, V) is successful, the tree examines the containment heat removal function (W). In this illustration, containment heat removal is limited to alternate shutdown cooling (SDC) using LPCI train D, the available SRVs, train D heat exchanger, and the required support systems such as emergency service water. Containment venting

---

* It was necessary to assume some changes to the configuration of this fire area in order to allow data from the Limerick fire PRA to be used for this illustration. Therefore, this analysis does not model the Limerick plant.

** Again, since the purpose of this analysis is to illustrate the process, the most readily available data were used to quantify the event tree.

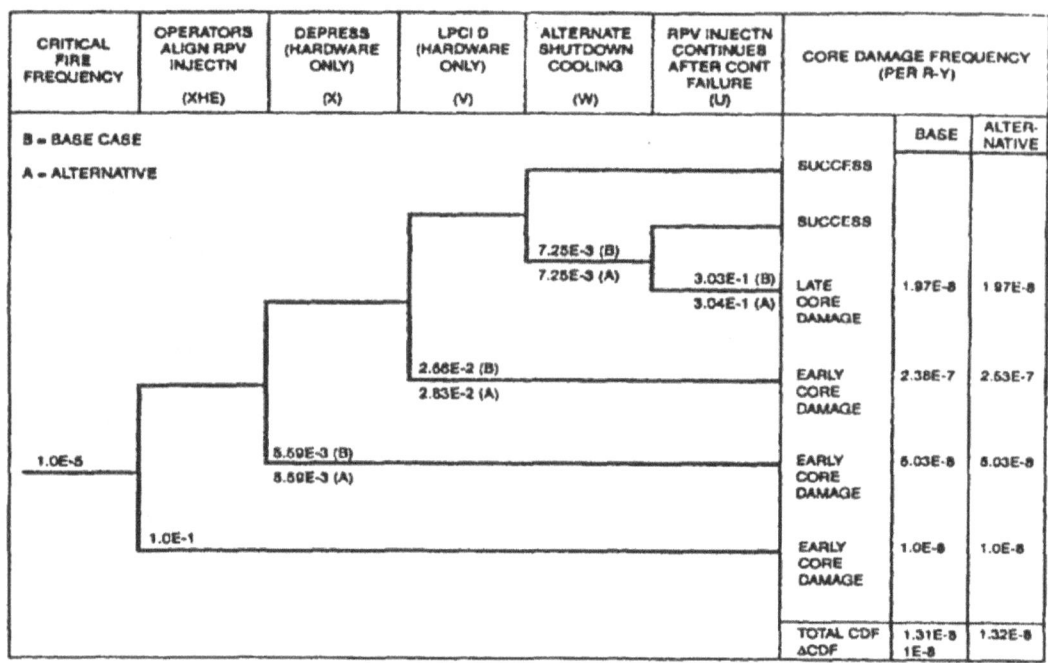

CRITICAL FIRE FREQUENCY	OPERATORS ALIGN RPV INJECTN (XHE)	DEPRESS (HARDWARE ONLY) (X)	LPCI D (HARDWARE ONLY) (V)	ALTERNATE SHUTDOWN COOLING (W)	RPV INJECTN CONTINUES AFTER CONT FAILURE (U)	CORE DAMAGE FREQUENCY (PER R-Y)		
							BASE	ALTER-NATIVE
B = BASE CASE						SUCCESS		
A = ALTERNATIVE						SUCCESS		
				7.25E-3 (B) 7.25E-3 (A)	3.03E-1 (B) 3.04E-1 (A)	LATE CORE DAMAGE	1.97E-8	1.97E-8
			2.66E-2 (B) 2.83E-2 (A)			EARLY CORE DAMAGE	2.38E-7	2.53E-7
1.0E-5		8.59E-3 (B) 8.59E-3 (A)				EARLY CORE DAMAGE	8.03E-8	8.03E-8
	1.0E-1					EARLY CORE DAMAGE	1.0E-6	1.0E-6
						TOTAL CDF ΔCDF	1.31E-8 1E-8	1.32E-8

Figure 6.11
Quantified Event Tree for Loss-of-Offsite-Power Case Study

is not available. The suppression pool will continue to heat up to boiling and pressurize the containment. At some point (in approximately 24 hours), the containment pressure will be sufficient to close the SRVs. The RPV will repressurize and fail LPCI. Continued RPV injection is available using the CRD hydraulic system. The containment will continue to pressurize and eventually rupture. The Peach Bottom analysis assumes that all injection systems taking suction from the suppression pool (including LPCI) fail after the containment ruptures. The simplified tree uses the top event to examine the operation of the CRD system before and after containment failure. (Although single-pump CRD injection is not sufficient early in the event, a plant-specific Peach Bottom analysis has shown that it will prevent core damage at about 8 hours after scram.) If CRD injection continues, core damage is averted and this is considered a success. A failure of CRD injection will result in late core damage (in more than 24 hours).

The estimated CDF for the base case is 1.31E-6 per reactor-year. The alternative case is slightly higher at 1.32E-6 per reactor-year because of the loss of ac power redundancy. The CDF difference is only 1.E-8 per reactor-year essentially due to the low probability for a LOSP. As shown in the event tree, these results are dominated by the 0.1 probability assumption for operator error XHE. The corresponding sequence accounts for about 75 percent of the total CDF for each case. Unlike the 72-hour-to-cold-shutdown case study, this human error is common to both the regulatory compliance case and the alternative approach. Similarly, the fire modeling, used to predict the extent of fire damage, is also used in both case studies. Although one could prejudge the LOSP assumption to be of marginal value because of the low probability of a LOSP, creating and quantifying an event tree allows one to methodically show its impact on dominant accident sequences, and on the overall CDF.

The preceding analysis illustrates how information from a fire PRA* can be used to

---

*Plant-specific applications will require accurate models of plant configurations and data representing plant conditions.

determine the value of assumptions required by current regulations by identifying the risk-significant fire scenarios in which the assumption may have value, and examining the impact of the assumption on risk-significant scenario development and quantification.

7 APPLICATION COST BENEFITS

Chapter 6 examined the insights that could be drawn useful to the regulatory process utilizing performance-based, risk-informed approaches for selected aspects of current prescriptive and deterministic requirements. In this chapter, these methods are evaluated to see if implementation would have economic benefits for licensees. This chapter only provides information to indicate the type and approximate amount of potential savings using these alternate approaches. It is not intended to provide a cost-benefit analysis to support any regulatory action.

Several case studies can be characterized as one-time events. These case studies are generally very plant specific and have limited industry-wide application. Other case studies show cost reductions for recurring costs, primarily surveillance. These alternate approaches are generally applicable to a large number of licensees.

Some of these case studies have been accommodated under the existing regulations, i.e., through the exemption process, as a deviation, or as a safety evaluation under the Generic Letter 86-10 license condition. Other alternate approaches, such as the application addressing the loss-of-offsite-power requirements for alternate or dedicated shutdown, do not appear to have been implemented under the current regulatory framework.

The estimated costs of the technical evaluations are adapted from information developed by the NRC to estimate licensee and NRC costs for technical specification changes (NUREG/CR-4627).

The estimated licensee costs are $18,000 for straightforward technical specification changes and $35,000 for more complex revisions. These estimates are based on 8 and 16 staff weeks of utility technical, legal, management, and committee input at $55 per staff hour in 1988 dollars. Total costs are rounded to the nearest thousand dollars. Assuming a 1995 professional staff rate of $76 per hour* yields estimated licensee costs of $24,000 and $49,000 for straightforward and complex technical specification changes, respectively. These estimates consist of nontechnical and technical components. The nontechnical contribution includes licensing effort, upper management review, and support to the NRC review process (e.g., meetings and submittals of additional information). The technical scope (and cost) of a technical specifications change is considered to be equivalent to the licensee's technical evaluation for an issue of the same complexity.

The case studies and licensee initiatives presented in Chapter 6 require varying levels of technical effort. In recognition, the estimated licensee technical levels of effort for these regulatory alternatives have been subdivided into three levels: straightforward, complex, and very complex technical evaluations.

Straightforward evaluations require a limited amount of technical input. The major technical effort might consist of determining the plant-specific licensing bases with regard to a specific regulatory requirement. An example is the battery capacity testing application discussed in Chapter 6. On the basis of estimates, the technical effort associated with each of these examples is less than 2 weeks each, or $3,000–$5,000. The licensee's cost to process a straightforward technical specification change is $24,000, as developed above. By extension, this assumes that the exemption process entails a cost of ~$20,000 for nontechnical support, i.e., a certain minimum level of licensing effort is required, regardless of the issue's complexity.

Complex evaluations require more significant technical input. The fire detector surveillance application, which develops a plant-specific reliability database, is one example of a complex

---

* Inflated to 1995 dollars assuming wages kept pace with the long-term forecast for inflation of 4.8 percent per year.

evaluation. The licensee's cost to develop and support a complex technical specification change ($49,000) has been developed above. The nontechnical level of effort is again assumed to be $20,000. The technical cost is, therefore, $29,000. The technical effort takes about 4 staff months (60 percent of the total). This is believed to be reasonable for this level of complexity.

A third category, the very complex evaluation, has been defined to account for those issues that require state-of-the-art probabilistic risk assessment (PRA) or fire modeling analyses. In recognition of the significant effort required, the estimated technical cost has been increased by a factor of 2 to $58,000.

The regulatory compliance requirements and the associated costs are not addressed here. They are additional plant-specific variables that could somewhat reduce the cost savings developed below.

7.1 EMERGENCY LIGHTING

The premature failure of 8-hour safe-shutdown emergency lighting as a result of full discharge surveillance testing was discussed in Section 6.2.1.2.2. That section developed a reliability approach for testing 8-hour-rated battery packs for 5 hours to avoid this type of failure.

A second approach to this issue is being considered at the Catawba plant, as discussed in Section 7.6. The design basis for each light was reviewed. In most plant areas, battery-operated safe-shutdown emergency lighting is not required for an 8-hour duration. For example, the lights that illuminate the path from the main control room (MCR) to the remote shutdown, are required in the first few minutes after a fire in the control room. Once the operators have evacuated the MCR, these lights have fulfilled their design function. Duke Power Company, the Catawba licensee, is examining the feasibility of redesignating the emergency lighting that is only required for the short term (i.e., less than 30 minutes) as 1-hour lighting. These lights would continue to have 8-hour-rated batteries; however, the annual capacity testing would be for 1 hour. This would assure that these emergency lights

satisfied their design bases while avoiding battery capacity degradation caused by a full-discharge-capacity test.

The approximate cost savings that could ensue from changing the 8-hour regulatory test requirements is developed below. It is a combination of the reliability projections of Section 6.2.1.2.2 and the plant-specific information from Catawba provided by Duke Power Company.

The base case (T) develops a cost estimate for emergency lighting replacement for the current regulatory requirement. A 40-percent battery failure rate* is assumed to be incurred during the yearly testing and maintenance. The failure of a battery is defined as failing to satisfy the 8-hour-rating requirement. The time required for the 8-hour-capacity test is assumed to be about 3 hours per emergency lighting unit.*

The alternative case (T′) assumes that a 1-hour-capacity test is appropriate for the majority (80 percent)* of the safe-shutdown lights. The rest of the lights continue to be required for a full 8-hour duration; however, they are tested for 5 hours. In accordance with Section 6.2.1.2.2, the failure rate is expected to be reduced to an equilibrium failure rate of about 15 percent per year.

The time required for 1-hour-capacity testing is assumed to be 1 hour. The time required for the 5-hour-capacity test is conservatively assumed to require 3 hours, i.e., the same as an 8-hour-capacity test.

The following parameters are assumed for both cases:

- Fifty safe-shutdown emergency lights are installed in the unit*

- The cost of labor is $43 per hour for technicians.*

- Replacement batteries cost $100 each.*

---

*Catawba plant-specific values

- Four hours of labor are assumed for replacing each failed battery.*

The cost estimate for the base case is the sum of three components:

C1: The capital cost for battery replacement

$$C1 = 50 \text{ lights/reactor} \times 40\% \text{ failure rate/year} \times \$100/\text{battery}$$
$$= \$2,000/\text{reactor-year}.$$

C2: The labor cost associated with unit troubleshooting and battery replacement

$$C2 = 50 \text{ lights/reactor} \times 40\% \text{ failure rate/year} \times 4 \text{ hours/failure} \times \$43/\text{hour}$$
$$= \$3,440/\text{reactor-year}$$

C3: The annual cost associated with the 8-hour-capacity test

$$C3 = 50 \text{ lights/reactor} \times 3 \text{ hours/light/year} \times \$43/\text{hour}$$
$$= \$6,450/\text{reactor-year}$$

The total estimated annual cost of the present 8-hour-test requirement for emergency lighting is

$$T = C1 + C2 + C3$$
$$= \$11,890/\text{reactor/year}$$

Similarly, the cost estimate for the alternative case can be calculated as

$$C1' = 50 \text{ lights/reactor} \times 15\% \text{ failure rate/year} \times \$100/\text{battery}$$
$$= \$750/\text{reactor-year}$$

$$C2' = 50 \text{ lights/reactor} \times 20\% \text{ failure rate/year} \times 4 \text{ hours/failure} \times \$43/\text{hour}$$
$$= \$1,720/\text{reactor-year}$$

*Catawba plant-specific values

C3': 80 percent of the lights (40 lights) are redesignated as 1-hour-rated lights. The 8-hour-capacity batteries are tested for 1 hour. The remaining 20 percent of the lights (10 lights) retain their 8-hour requirement and are tested for 5 hours

$$C3' = (40 \text{ lights} \times 1 \text{ hour/light/year} \times \$43/\text{hour}) + (10 \text{ lights} \times 3 \text{ hours/light/year} \times \$43/\text{hour})$$
$$= \$3,010/\text{reactor-year}$$

and
$$T' = C1' + C2' + C3'$$
$$= \$5,480/\text{reactor-year}$$

The projected annual savings is

$$T - T' = \$11,890 - \$5,480$$
$$= \$6,410/\text{reactor/year}$$

Assuming a remaining life of 20 years for this unit, and constant annual savings, the present value of the savings is about $55,000 per reactor (10-percent discount rate) and $80,000 per reactor at a 5-percent discount rate.

Cost

The Catawba licensee estimated that the engineering effort that was required to investigate the design bases of each safe-shutdown emergency light and revise affected documents and procedures totaled about $3,000.

This results in an estimated net savings per reactor of $52,000 to $77,000 (at 10-percent and 5-percent discount rates, respectively).

7.2 THE 72-HOUR CRITERION TO REACH COLD SHUTDOWN

Section 6.2.2.2.1 examines an alternate approach to the Appendix R requirement to reach cold shutdown in 72 hours. This application presents a methodology that can be used to evaluate the risk impact of a protracted time to reach cold shutdown. Selected exemptions from this Appendix R requirement have been granted in the past, so the economic value of this approach may be limited to the avoidance of the expenses of a formal exemption request. However, at the other extreme, if this methodology can provide justification that a conforming plant modification is not risk warranted, then the avoided cost can be substantial.

Like most nuclear plant modifications, the costs associated with this application to ensure compliance with the 72-hour cold-shutdown requirement are highly plant specific. The application postulates extensive damage to both trains of the residual heat removal (RHR) system that cannot be repaired within 72 hours. This implies a degree of fire damage that is not limited to cabling. Major components must be protected. For the purposes of Appendix R compliance, the costs associated with the installation of a 3-hour-rated fire barrier were examined. The wall is assumed to be 4.6 m (15 ft) high and 6.1 m (20 ft) wide, bisect the fire area, and separate the two RHR trains. A 1982 Sandia report (Dube, 1982) provided an estimate that has been modified to account for inflation (at 5 percent per year) and a factor of 2 to account for the additional seismic design and construction costs. For this case, the estimated cost of the fire wall construction is about $160,000 in 1995 dollars.

However, this is not the entire cost of the modification. Not considered are the costs associated with equipment relocation to accommodate the fire wall; potential heating, ventilation, and air conditioning modifications; fire wall penetration protection for piping, cabling, ducting, and doors; and replacement power. The latter consideration can easily dominate the total cost if the installation extends plant outage. In addition to the capital costs, this modification would incur periodic surveillance and maintenance costs for the fire barrier penetrations, dampers, and doors.

The technical evaluation for this application consisted of extending a plant PRA to accommodate a 200-hour mission time. A plant-specific technical evaluation would also examine plant capability, system interlocks, procedures, and operator action. This is considered to be a very complex technical evaluation. The cost of this evaluation is $58,000, as discussed at the beginning of this chapter. The net avoided cost is, therefore, $160,000–$58,000, or $102,000.

This chapter does not purport to present exact costs associated with compliance with the 72-hour criterion of Appendix R. The costs are too plant specific. Rather, the intent is to convey that the costs can be significant for this situation.

7.3 COST EVALUATION OF FIRE DETECTOR CASE

Section 6.2.1.2.1 discusses a study by SINTEF that looks at the feasibility of adopting a performance-based surveillance testing approach for fire detectors. This section examines the approximate cost saving that could be realized from a change from an annual detector surveillance (the base case) to a 10-percent rotating sampling annual test interval (the alternative case).

The following parameters are assumed for both cases:

- Approximately 2,000 detectors are installed in the plant.

- Ten minutes is required for each detector for the surveillance.

- The cost of labor is $43 per hour for technicians.

The cost estimate for the base case (T) is:

$$
\begin{aligned}
T \quad = \quad & 2,000 \quad \text{detectors/reactor} \\
& \times \ 10 \quad \text{minutes/detector-year} \\
& \times \ \$43/\text{hour} \\
= \quad & \$14,333/\text{reactor-year}
\end{aligned}
$$

The alternate case, T′ is:

$$
\begin{aligned}
T' \quad = \quad & 2,000 \ \text{detectors/reactor} \quad \times \ 10\text{-} \\
& \text{percent} \quad \text{sample} \\
& \times \ 10 \quad \text{minutes/detector-year} \\
& \times \ \$43/\text{hour} \\
= \quad & \$1,433/\text{reactor-year}
\end{aligned}
$$

The estimated annual savings is:

$$
\begin{aligned}
T - T' \quad = \quad & \$14,333 - \$1433 \\
= \quad & \$12,900/\text{reactor/year}
\end{aligned}
$$

Assuming a remaining life of 20 years for this unit and constant annual savings, the present value of the savings is about $110,000 per reactor (10-percent discount rate) and $161,000 per reactor at a 5-percent discount rate.

Cost

This is considered to be a complex technical evaluation. The licensee's cost is estimated at $29,000, as discussed above.

Therefore, the net value of the savings ranges from $81,000 (10-percent discount) to $132,000 (5-percent discount) per reactor.

Please note, this estimate does not develop a projected detector failure rate for the purposes of this cost estimate. It assumes the detector reliability target is readily attainable, i.e., no failures are anticipated.

7.4 SAFE SEPARATION DISTANCE

The safe separation analysis of Section 6.2.1.3 presents a performance-based and risk-informed approach to examine departures from the current regulatory requirements. The avoided cost of this approach can range from the incremental cost of a formal exemption to the cost associated with physical plant modifications. Several licensees submitted cost estimates for modifications to ensure prescriptive compliance. The engineering and installation cost for the modifications cited were estimated at $420,000 and $3,350,000, respectively. Lost revenue was estimated at $24 million if immediate installation was required.

For the purposes of this case study, regulatory compliance assumes that the target cable trays are wrapped with 1-hour-rated fire blankets. The cost of material, labor, and installations for this modification is estimated to be about $1,500 per foot of cable tray, or $225,000 total. Other factors, such as seismic reanalysis or the need for a forced outage, are not considered. The net cost savings for this modification is the avoided cost of the modification as reduced by the cost of the very complex technical evaluation ($58,000), or about $167,000.

7.5 THE LOSS-OF-OFFSITE POWER REQUIREMENT FOR ALTERNATIVE OR DEDICATED SHUTDOWN CAPABILITY

The development of plant-specific individual plant examinations of externally initiated events (IPEEEs) in conjunction with the refinement of fire modeling capabilities has enabled licensees to predict the consequences of a fire in a particular area. Not all of the fires in areas that require alternative shutdown capability would induce a loss of offsite power (LOSP). This section examines the potential cost savings that could be associated with a performance-based approach to this requirement. Although the fire protection requirements have been implemented and any necessary plant modifications have been completed, additional nonconformances may occasionally arise as a result of an inspection or a licensee self audit.

If, in the future, a licensee determines that a scenario requires alternative shutdown capability, the approach of Section 6.2.2.2.2 can be used to determine if a fire-induced LOSP is likely.

If the LOSP is limited to random, independent events, a case can be made that the protection of one train of emergency power is not necessary. For the purposes of the case study in Section 6.2.2.2, 6.1 m (20 ft) of cable tray wrapping is required.

This modification assumes that 6.1 m (20 ft) of cable tray are wrapped at a cost of $30,000. The cost of seismic reanalysis or derating is not considered. However, the technical evaluation to justify not protecting the cable tray is estimated to cost $58,000. This illustrates that for limited-scope modifications, a hardware fix may sometimes be more economical.

A more widespread application of this examination of the LOSP requirement is the potential to reduce the number of emergency lights. The Catawba plant, as a result of its IPEEE, has determined that the only fires that can induce a LOSP occur in the turbine building. Fires in safety areas do not cause loss of offsite power. As a consequence, that licensee has estimated that about 40 percent (or 20 lights) of each unit's safe-shutdown lights will never be demanded for any fire in an area that can affect safe-shutdown equipment. These lights generally illuminate the paths from the main control room to the auxiliary shutdown panel or the standby facility.

The licensee has estimated that the elimination of these 20 safe-shutdown lights would reduce the recurring costs associated with surveillance testing and the repair of the failed units.

The following additional parameters* are assumed for this cost evaluation:

- Forty percent of the safe-shutdown emergency lights fail the annual capacity test.

- The cost of labor is $43 per hour for technicians.

- Replacement batteries cost $100 each.

- The monthly surveillance test takes 8 minutes per emergency light.

- The annual capacity test takes 3 hours per light.

- Four hours of labor are assumed for replacing each failed battery.

The cost savings consists of the following avoided costs:

C1: The labor cost associated with the monthly surveillance

$$C1 = \text{20 lights/reactor}$$
$$\times \text{8 minutes/month}$$
$$\times \text{12 months/year}$$
$$\times \text{\$43/hour}$$

$$= \text{\$1,376/reactor-year}$$

C2: The labor cost for the annual capacity test

---

*These parameters are plant-specific values from Catawba

C2 = 20 lights/reactor
$$\times \text{3 hours/light/year}$$
$$\times \text{\$43/hour}$$

$$= \text{\$2,580/reactor-year}$$

C3: The capital cost of battery replacements

$$C3 = \text{20 lights/reactor}$$
$$\times \text{40-percent failure rate/year}$$
$$\times \text{\$100/battery}$$

$$= \text{\$800/reactor-year}$$

C4: The labor cost associated with unit troubleshooting and battery replacement

$$C4 = \text{20 lights/reactor}$$
$$\times \text{40-percent failure rate/year}$$
$$\times \text{4 hours/failure}$$
$$\times \text{\$43/hour}$$

$$= \text{\$1,376/reactor-year}$$

The total estimated annual savings is:

$$T = C1 + C2 + C3 + C4$$
$$= \text{\$6,132/reactor/year}$$

Assuming a remaining life of 20 years for this unit and assuming constant annual savings, the present value of the savings ranges between $52,000 (10-percent discount) and $76,000 (5-percent discount) per reactor.

Cost

The Catawba licensee evaluated this change as part of the $3,000 engineering effort to redesignate most of the safe-shutdown lights to 1-hour capacity. We will conservatively use the same cost for this effort.

This results in an estimated net savings of $49,000–$73,000 per reactor.

Please note that this cost savings is an independent estimate and does not credit an improvement in the annual battery capacity test failure rate that could be expected from a reduction in test duration (see Section 6.2.1.2.2).

This would reduce the present value of the savings due to avoided failures (C3 and C4) by more than 50 percent, to $38,000–$57,000 per reactor. Therefore, the total estimated cost savings (at 5-percent discount) if these two initiatives were implemented together is about $134,000 per reactor.

7.6 OTHER LICENSEE INITIATIVES

This section develops estimated cost savings for several initiatives by Duke Power Co (DPC). at the Catawba nuclear power station. These initiatives are discussed followed by an evaluation of the cost savings.

Fire Barriers

When Catawba was under design and construction, barriers were specified on a conservative basis. Since the plant has a dedicated safe-shutdown system, many of these barriers are not needed from a regulatory compliance perspective. The basis of each fire barrier in the Catawba site was recently reexamined.

Fire barriers were classified as not required, insurance, or NRC committed. The barriers that are "not required" are not necessary to meet NRC regulations. In addition, their placement does not allow these barriers to effectively limit the spread of fires. Approximately 80 barriers and 875 seals in each unit were re-designated for insurance (loss control) purposes or determined to not be required. The remaining barriers generally separate redundant analyzed safe-shutdown trains, separate the control complex from the rest of the plant, enclose high hazard areas (e.g., the switchgear room), separate safety from non safety areas. These barriers are designated as NRC-committed barriers. They remain in the fire protection program and continue to be subject to regular inspections and fire watches.

Smoke and Heat Detectors

Catawba was designed to the Duke standard at that time, which utilized smoke and heat detectors as companions. Duke subsequently realized that it had too many detectors. (The regulations require providing adequate detection and meeting NFPA requirements.) The inspection and testing requirements for these detectors, originally in the technical specifications, were moved to the selected license commitments section of the FSAR.

Each location was evaluated to determine which detector type would be most effective. Generally, the smoke detectors were retained. Approximately 350 detectors per unit were eliminated.

Each location still has one detector. The next phase of this effort will focus on the need for detection at each location. Detection in the plant is laid out on a 20 × 20 grid. Most of these detectors are not protecting redundant trains. DPC believes the existing plant layout exceeds the requirements of NFPA 72 and that some additional detectors/locations can be eliminated.

Fire Protection Valve Inspections

At Catawba, approximately 48 hours a month (for both units) are spent confirming fire protection system valve positions. About 400 valve sites are inspected. The valves are locked and under operations key control. In three years none of these valves has been found in the wrong position. Using the safety evaluation process, DPC has proposed increasing the surveillance interval based on past experience. The inspection interval would go to quarterly, semiannually, and finally to an annual basis if they maintain a greater than 99-percent success rate. This proposed change is presently being evaluated internally by the licensee.

Emergency Lighting

Each Catawba unit has about 150 emergency lights, 50 of which are safe-shutdown lights. The safe-shutdown units are installed on paths from the main control room to the auxiliary shutdown panel and the standby shutdown system (within the plant). Fires were postulated on these paths and alternate routes were also lighted. Because of the high ambient temperatures in many locations, the plant has experienced a significant number of failures during the annual 8-hour capacity test.

DPC has examined the design basis/purpose of each light. The FSAR, SER, BTP, Appendix R and the SBO rule were all reviewed. In general only short-term lighting is required, i.e., to permit passage through an area or isolate letdown paths. In addition, the IPEEE demonstrated that fires in safety areas did not induce losses of offsite power (LOOPs). Main generator fires were the primary cause of LOOP.

Most of Catawba's safe-shutdown emergency lights do not need to function for 8 hours and could be redesignated as 1- or 4-hour ratings. The majority of these lights are not specified in the SER and can be redesignated using the 50.59 process.

DPC believes this lighting redesignation would result in a significant cost savings without any safety impact.

Fire Extinguishers

When the Catawba fire protection plan was being developed, NFPA10 was used to determine fire extinguisher locations. Over time more extinguishers were added. At Catawba, about 6 staff days per month are expended for the monthly extinguisher surveillance required by NFPA10. (At Oconee and McGuire this surveillance takes much longer because those plants aren't bar coded). DPC reexamined the basis for each extinguisher and their regulatory commitment (NFPA10). Duke established that 80 (or approximately 25 percent) of the extinguishers could be removed from each unit without violating the licensing commitment. Once again the safety evaluation process can be used to delete most of these extinguishers. The extinguishers that are credited in the SER would require an exemption request, however.

A second phase of this effort would increase the surveillance frequency of the extinguishers that remain in the plant. Since NFPA 10 is part of the licensing basis and it specifies surveillance requirements, a license amendment may be required to institute this proposed change.

7.6.1 Fire Barriers

When Catawba was under design and construction, fire barriers were specified on a conservative basis. The design bases of each fire barrier in the plant were reexamined recently.

Approximately 80 barriers and 875 penetration seals in each unit were not required to ensure compliance with NRC fire protection regulations. These barriers and seals were redesignated and are no longer subject to regular inspections or fire watches. The licensee examined plant records, before and after the redesignation, to estimate these cost savings.

C1: Monthly Inspection Time
About 5 percent of the barriers are inspected each month to ensure that all barriers are checked once every 18 months.

Approximately 25 hours per month are being saved on the fire barrier and penetration seal inspections.

$$C1 = 25 \text{ hours/reactor month}$$
$$\times 12 \text{ months/year}$$
$$\times \$43/\text{hour}$$

$$= \$12,900/\text{reactor-year}$$

C2: Fire Watches
In 1990, prior to the barrier redesignation, each unit had about 260 fire watches.

Approximately 150 fewer fire watches per year are required after the redesignation. This is a 58-percent reduction. An average time of 3 hours per fire watch is assumed.

$$C2 = 150 \text{ fire watches/year}$$
$$\times 3 \text{ hours/fire watch}$$
$$\times \$43/\text{hour}$$

$$= \$19,350/\text{reactor-year}$$

C3: Barrier and Seal Repairs
Before the barrier redesignation, about 18 repairs per year were required because of

inadequacies discovered during the surveillance. Repairs were estimated to cost $200 each. The savings in repairs for the redesignated barriers can be estimated on the basis of the reduction in fire watches.

$C3$ = 58-percent reduction in fire watches
× 18 repairs/year
× $200/repair

= $2,088/reactor-year

$C4$: Anti-Contamination Clothing
Many of the fire barriers are in radiological control zones (RCZs) that require the use of "anti-Cs." This redesignation initiative eliminated the need to go into several RCZs for barrier and seal surveillance. Eight sets of anti-Cs are saved. Avoided dose, dressout time, and radwaste disposal costs are not considered.

$C4$ = 8 sets/year × $30/set
= $240/reactor-year

The annual savings, T, is:

T = $12,900 + $19,350 + $2,088
+ $240

= $34,578/reactor-year

Assuming a remaining life of 20 years and a constant annual savings, yields a present value per reactor of $294,000 (10-percent discount rate) to $431,000 (5-percent discount rate).

Cost

The licensee used the 10 CFR 50.59 process to redesignate the fire barriers. The effort was a minor modification and was estimated to cost about $5,000.

Therefore, the projected net savings is about $289,000–$426,000 per reactor, depending on the discount assumption.

7.6.2 Smoke and Heat Detectors

Catawba was designed to the Duke Power Company standard which, at that time, specified smoke and heat detectors as companions. Each location was estimated to determine which detector type would be most effective. Approximately 350 detectors in each unit were eliminated. The licensee estimated a modest time savings of 1 minute per detector for the semiannual visual inspection and 10 minutes for the 18-month bench testing of each detector. This is a savings of about 50 hours per year (or about $2,150 per reactor-year). For a 20-year remaining life and constant annual savings, the present value per reactor is $18,000 (10-percent discount) to $27,000 (5-percent discount). The licensee estimated that the effort to implement this change was about a week, or $3,800 at the $95 per hour rate for engineering. This results in a net savings of $14,000–$23,000 per reactor.

7.6.3 Fire Protection Valve Inspections

Catawba is evaluating a performance-based inspection methodology for the fire protection system valve alignments. The inspection would eventually reach an annual interval if more than a 99-percent success rate was maintained. Catawba presently expends 24 hours per unit for the monthly valve inspection. The reduction in the annual surveillance interval is projected to save $11,352 per reactor-year (11 inspections/year × 24 hours/inspection × $43/hour).

The cost for this proposed change was estimated to be about $3,000. The yearly trending cost was neglected in this evaluation. The net lifetime savings is $94,000–$138,000 at 10-percent and 5-percent discount rates.

7.6.4 Emergency Lighting

The emergency lighting initiatives at Catawba were integrated into the emergency lighting surveillance case study discussed earlier.

7.6.5 Fire Extinguishers

Removal of Selected Extinguishers

The basis for each fire extinguisher at the Catawba plant was recently reviewed. The licensee found more fire extinguishers than required by its regulatory commitments. Of the approximately 230 extinguishers per unit, 80 can be removed. This would result in an annual cost savings attributable to avoided surveillance and maintenance costs. Although Duke Power intends to use these 80 extinguishers elsewhere in its system, this salvage value has been conservatively neglected. The licensee estimated that the monthly surveillance takes 20 minutes (0.33 hour) for each extinguisher and the annual maintenance costs $20 each.

C1: Monthly Inspections
$$C1 = 80 \text{ extinguishers removed/reactor} \times 0.33 \text{ hour/extinguisher/month} \times 12 \text{ months/year} \times \$43/\text{hour}$$

$$= \$13,622/\text{reactor-year}$$

C2: Annual Maintenance
$$C2 = 80 \text{ extinguishers/reactor} \times \$20/\text{extinguisher/year}$$

$$= \$1,600/\text{reactor-year}$$

The annual cost savings is:

$$\$13,622 + \$1,600$$
$$= \$15,222 \text{ per reactor-year.}^*$$

The present value of these savings is $130,000 (10-percent discount rate) to $190,000 (5-percent discount rate) per reactor.

---

* Like the fire barrier initiative, this effort has also reduced the number of RCZ entries for surveillance; however, the savings are neglected for this evaluation.

Cost

The safety evaluation process was used to examine the impact of removing these fire extinguishers. The licensee has estimated that the total cost to implement this change will be about $2,000.

The estimated total savings (T) for this effort is

$$T = (\$130,000 - \$190,000) - \$2,000$$
$$= \$128,000 - \$188,000/\text{reactor}$$

Performance-Based Surveillance Initiative

Duke Power is also examining the feasibility of instituting a performance-based surveillance program to replace the current monthly surveillance requirement. Duke anticipates extending this surveillance to semiannually. On the basis of about 230 extinguishers remaining in each unit, the following costs are noted:

- Current Inspection Cost
$$C = 230 \text{ extinguishers/reactor} \times 0.33 \text{ hour/extinguisher surveillance} \times 12 \text{ surveillances/year} \times \$43/\text{hour}$$

$$= \$39,165/\text{reactor-year}$$

- Projected Inspection Cost
$$C' = 230 \text{ extinguishers/unit} \times 0.33 \text{ hour/extinguisher surveillance} \times 2 \text{ surveillances/year} \times \$43/\text{hour}$$

$$= \$6,528/\text{reactor-year}$$

- Annual Cost Savings
$$C - C' = \$39,165 - \$6,528$$
$$= \$32,637/\text{reactor-year}$$

Assuming a remaining reactor life of 20 years and an implementation cost of $4,000, the present value of the net cost savings is approximately $275,000–$403,000 at 10-percent and 5-percent discount rates, respectively.

8 REFERENCES

Alpert, R.L., and E.J. Ward, "Evaluating Unsprinklered Fire Hazards," Factory Mutual Research Corp., Norwood, Massachusetts, August 1982.

American National Standards Institute/Institute of Electrical and Electronics Engineers, ANSI/IEEE 450-1987, "IEEE Recommended Practice for Maintenance, Testing, and Replacement of Large Lead Storage Batteries for Generating Stations and Substations," New York, 1987.

————, ANSI/IEEE 485-1983, "IEEE Recommended Practice for Sizing Large Lead Storage Batteries for Generating Stations and Substations," New York, July 1984.

American Society of Heating, Refrigeration, and Air Conditioning Engineers, "Design of Smoke Management Systems," Atlanta, Georgia, 1993.

Azarm, M.A., "Impacts of Underreported and Event Screening on Fire Frequencies," Brookhaven National Laboratory Technical Report JCN W-6276, Upton, New York, February 1998.

Azarm, M.A., and T.L. Chu, "On Combining the Generic Failure Rate Data for Probabilistic Risk Assessment," pp. 1525–1529 in *Probabilistic Safety Assessment and Management*, Vol. 2, Elsevier Science Publishing Co., Inc., New York, 1991.

Beard, A., "Evaluation of Deterministic Fire Models," *Fire Safety Journal*, Vol. 19, pp. 295–306, 1992.

Bento, J.P., et al., "Reliability Data Book for Components in Swedish Nuclear Power Plants," Nuclear Safety Board of Swedish Utilities, Stockholm, 1985.

Bertrand, R., et al., "Studies of Fire Development—Development of FLAMME-S Computer Code," Proceedings of ICONE 5: Nuclear Advances Through Global Cooperation, Nice, France, May 1996.

British Standards Institution, "British Standard Code of Practice for the Application of Fire Safety Engineering Principles to Fire Safety in Buildings," 94/340340 DC, draft, London, 1994.

————, "Fire Safety Engineering in Buildings: Guide to the Application of Fire Safety Engineering Principles," DD 240: Part 1, London, 1997.

————, "Fire Safety Engineering in Buildings: Commentary on the Equations Given in Part 1," DD 240: Part 2, London, 1997.

Bruce, G.E., "Performance Based Fire Detector Surveillance Testing—A Nuclear Plant Application," American Nuclear Society International Topical Meeting, *Proceedings on Safety of Operating Reactors*, September 17–20, Seattle, Washington, 1995.

Clarke, F.B., R.W. Bukowski, S.W. Steifel, J.R. Hall, and S.A. Steele, "The National Fire Risk Assessment Research Project Final Report," National Fire Protection Research Foundation, Batterymarch Park, Quincy, Massachusetts, 1990.

Code of Federal Regulations, Title 10, "Energy," U.S. Government Printing Office, Washington, D.C., revised periodically.

Commonwealth Edison Co., "Zion Station Unit 1 and 2 Probabilistic Safety Study," NRC Docket Nos. 50-295 & 50-304, Zion, Illinois, September 1981.

Consolidated Edison Company of New York, Inc./Halliburton NUS Environmental Corp., "Individual Plant Examination for Indian Point 2 Nuclear Generating Station," Buchanan, New York, August 1992.

Consumers Power Company, "Big Rock Point Plant Probabilistic Risk Assessment," Jackson, Michigan, March 1981.

Cross, R.B., et al., "South Texas Project Electric Generating Station Level 2 PRA and Individual

Plant Examination," Houston Lighting & Power Company and Pickard, Lowe, and Garrick, Palacios, Texas, August 1992.

Custer, Richard L.P., and Brian J. Meacham, "Introduction to Performance-Based Fire Safety," Society of Fire Protection Engineers and National Fire Protection Association, Quincy, Massachusetts, 1997.

Davis, W.D., and K. Notarianni, "NASA Fire Detection Study," NISTIR 5798, National Institute of Standards and Technology, Gaithersburg, Maryland, March 1996a.

Davis, W.D., et al., "Comparison of Fire Model Predictions With Experiments Conducted in a Hangar With a 15 Meter Ceiling," NISTIR 5927, National Institute of Standards and Technology, Gaithersburg, Maryland, 1996b.

Deal, S., "Technical Reference Guide for FPETOOL Version 3.2," NISTIR 5486-1, National Institute of Standards and Technology, Gaithersburg, Maryland, 1995.

———, "A Review of Four Compartment Fires With Four Compartment Fire Models," pp. 33–51 in *Fire Safety Developments and Testing, Proceedings of the Annual Meeting of the Fire Retardant Chemicals Association,* 1990.

Dube, R.A., "A Systematic Approach to the Identification and Protection From Fire of Vital Areas in Nuclear Power Plants," SAND 82-0648, Sandia National Laboratory, Albuquerque, New Mexico, October 1982.

Duke Power Company, "Catawba Nuclear Station IPE Submittal Report," Clover, South Carolina, September 1992.

Duong, D.Q., "Accuracy of Computer Fire Models: Some Comparisons With Experimental Data From Australia," *Fire Safety Journal,* 16: 6, pp. 415–431, 1990.

Electric Power Research Institute, EPRI NP-2660, "Fire Tests in Ventilated Rooms: Extinguishment of Fire in Grouped Cable Trays," Palo Alto, California, December 1992.

———, "IPEEE Submittal Report for McGuire Nuclear Station," Cornelius, North Carolina, November 1991.

———, EPRI NP-2751, "Fire Tests in Ventilated Rooms: Detection of Cable Tray and Exposure Fires," J.S. Newman, Palo Alto, California, 1983.

———, EPRI NP-7282, "COMPBRN IIIe: An Interactive Computer Code for Fire Risk Analysis," V. Ho et al., University of California at Los Angeles, May 1991.

———, EPRI TR-100370, "Fire-Induced Vulnerability Evaluation (FIVE)," Palo Alto, California, April 1992.

———, EPRI TR-100443, "Methods of Quantitative Fire Hazard Analysis," Palo Alto, California, 1992.

———, EPRI TR-105928, "Fire PRA Implementation Guide," December 1995.

Florida Power & Light Co., "Turkey Point Plant Units 3 & 4 Probabilistic Risk Assessment Individual Plant Examination," Final Report, Florida City, Florida, June 1991.

Friedman, R., "International Survey of Computer Models of Fire and Smoke." *Journal of Fire Protection Engineering,* Vol. 4, pp. 81–92, 1992.

Gallucci, R., and R. Hockenbury, "Fire-Induced Loss of Nuclear Power Plant Safety Functions," *Nuclear Engineering and Design,* Vol. 64, pp. 135–147, 1981.

Garrick, B.J., et al., "Seabrook Station Probabilistic Safety Assessment," Pickard, Lowe, and Garrick, Inc., Framingham, Massachusetts, December 1983.

Grand, A. F., ed., "Fire Standards in the International Marketplace," ASTM STP 1163, pp. 66–67, American Society of Testing and Materials, Philadelphia, Pennsylvania, 1995.

Ho, V., et al., "COMPBRN III, A Fire Hazard Model for Risk Analysis," *Fire Safety Journal,* Vol. 13, pp. 137–154, 1988.

Hokstad, P., et al., "A Reliability Model for Optimization of Schemes for Fire and Gas Detectors," *Reliability Engineering and System Safety*, Vol. 47, pp. 15–26, 1995.

Institute of Electrical and Electronics Engineers, IEEE 446, "IEEE Recommended Practice for Emergency and Standby Power System for Industrial and Commercial Applications," 1987.

Jones, W.W., "Progress Report on Fire Modeling and Validation," NISTIR 5835, National Institute of Standards and Technology, Gaithersburg, Maryland, May 1996

Kaplan, S., "On a Two Stage Bayesian Procedure for Determining Failure Rate From Experimental Data," *IEEE Transactions on Power Apparatus and Systems*, Vol. PAA-102, No. 1, January 1983.

Keski-Rahkonen, O., and S. Hostikke, "CIB W14 Round-Robin of Code Assessment: Design Report of Scenario A," VTT Technical Research Centre of Finland, Espoo, 1995.

Lambright, J., et al., "A Review of Fire PRA Re-quantification Studies Reported in NSAC/181," Draft Technical Report, Sandia National Laboratories, Albuquerque, New Mexico, April 1994.

Levin, J.P., and L.L. Kanz, "Fire Protection Database, FIREDAT User's Manual," Scientech Inc., Rockville, Maryland, January 1995.

Malhotra, H.L., "Fire Safety in Buildings," Building Research Establishment, Gerston, United Kingdom, 1987.

Martinez-Guridi, G., and M.A. Azarm, "Reliability Assessment of Electrical Power Supply to Onsite Class 1E Buses at Nuclear Power Plants," Technical Report FIN L-2505, Brookhaven National Laboratory, Upton, New York, June 7, 1994.

McGrattan, K.B., A. Hamins, and D.W. Stroup, "Sprinkler, Vent and Draft Curtain Interaction—Experiment and Computer Simulation," *Proceedings, Second International Conference on Fire Research and Engineering*, Society of Fire Protection Engineers, Gaithersburg, Maryland, 1997a.

McGrattan, K.B., Baum, H.R., Walton, W.D., and Trelles, J., "Smoke Plume Trajectory From In Situ Burning of Crude Oil in Alaska: Field Experiments and Modeling of Complex Terrain," NISTIR 5958, National Institute of Standards and Technology, Gaithersburg, Maryland, 1997b.

Meacham, Brian J., "The Evolution of Performance-Based Codes and Fire Safety Design Methods," Society of Fire Protection Engineers, Boston, Massachusetts, 1996.

Medford, M.O., Southern California Edison Company, letter to the NRC, "Docket Nos. 50-361 and 50-362, San Onofre Nuclear Generating Station, Units 2 and 3," March 13, 1987.

Mills, L.M., Tennessee Valley Authority, letter to H.R. Denton, "Docket Nos. 50-259, 260, 261," March 27, 1984.

Mowrer, F.W., and B. Gautier, "Comparison of Fire Model Features and Computations," *Proceedings, 17th Post-SMIRT Conference*, Lyons, France, 1997.

Mowrer, F.W., and D.W. Stroup, "Features, Limitations and Uncertainties in Enclosure Fire Hazard Analyses—Preliminary Review," NISTIR 5973, National Institute of Standards and Technology, Gaithersburg, Maryland, March 1998.

Nakaya, I., and Y. Hirano, "Japan's Approach Toward the Building Code and Standards of a New Generation," *Proceedings of the International Conference on Performance-Based Codes and Fire Safety Design Methods, Ottawa, Canada, September 1996*, Society of Fire Protection Engineers, Bethesda, Maryland, 1998.

National Fire Protection Association, *Fire Protection Handbook*, Seventeenth Edition, pp. 6-76, 6-77, Quincy, Massachusetts, July 1991.

———, National Codes, NFPA 10, "Portable Fire Extinguishers," Vol. 1, Quincy, Massachusetts, 1997.

———, NFPA 12, "Carbon Dioxide Extinguishing Systems," Vol. 1, Quincy, Massachusetts, 1997.

———, NFPA 13, "Sprinkler Systems, Installation," Vol. 1. Quincy, Massachusetts, 1997.

———, NFPA 25, "Inspection, Testing and Maintenance of Water-Based Fire Protection Systems," Vol.2, Quincy, Massachusetts, 1997.

———, NFPA 72, "National Fire Alarm Code," Vol. 4, Quincy, Massachusetts, 1997.

———, NFPA 72E, "Automatic Fire Detectors," Vol. 3, Quincy, Massachusetts, 1991.

———, NFPA 92B, "Smoke Management Systems in Halls, Atria and Large Areas," Vol. 10, Quincy, Massachusetts, 1997.

———, NFPA 101, "Life Safety Code," Vol. 5, Quincy, Massachusetts, 1997.

———, NFPA 1962, "Care, Use and Service Testing of Fire Hose Including Couplings and Nozzles," Vol. 9, Quincy, Massachusetts, 1997.

Nuclear Safety Analysis Center, "A Probabilistic Risk Assessment of Oconee Unit 3," NSAC 60, Vol. 1, W. R. Sugnet, G. J. Boyd, and S. R. Lewis, Electric Power Research Institute, Palo Alto, California, June 1984.

Nuclear Safety Commission, "Guide for Fire Protection in Light Water Nuclear Power Reactor Facilities," Tokyo, Japan, 1980.

NUS Corporation, "Severe Accident Risk Assessment, Limerick Generating Station," Pottstown, Pennsylvania, April 1983.

Parkinson, W., et al., "Fire Event Data Base for U.S. Nuclear Power Plants," NSAC-178-L/1992, Electric Power Research Institute, Palo Alto, California, 1992.

———, "Fire PRA Re-quantification Studies," NSAC-181, Electric Power Research Institute, Palo Alto, California, March 1993.

Peacock, R.D., S. Davis, and B.T. Lee, "An Experimental Data Set for the Accuracy Assessment of Room Fire Models, NBSIR 88-3752, National Bureau of Standards, Gaithersburg, Maryland, 1988.

Peacock, R.D., W.W. Jones, and R.W. Bukowski, "Verification of a Model of Fire and Smoke Transport," *Fire Safety Journal*, Vol. 21, pp. 89-129, 1993a.

Peacock, R.D., et al., "CFAST, the Consolidated Model of Fire Growth and Smoke Transport," NIST Technical Note 1299, National Institute of Standards and Technology, Gaithersburg, Maryland, February 1993b.

Peacock, R.D., et al., "A User's Guide for FAST: Engineering Tools for Estimating Fire Growth and Smoke Transport," Special Publication 921, National Institute of Standards and Technology, October 1997.

Pennsylvania Power & Light Company, "Susquehanna Steam Electric Station Individual Plant Examination for External Events," Vol. 2, 1994.

Portier, R.W., "Fire Data Management System, FDMS 2.0, Technical Documentation," NIST Technical Note 1407, National Institute of Standards and Technology, Gaithersburg, Maryland, February 1994.

Portier, R.W., R.D. Peacock, and W.W. Jones, "A Prototype FDMS Database for Model Verification," NISTIR 5835, National Institute of Standards and Technology, Gaithersburg, Maryland, May 1996.

Rackliffe, C.A., "Performance-Based Building Codes, the United Kingdom Experience," *Proceedings of the, International Conference on Performance-Based Codes and Fire Safety Design Methods, Ottawa, Canada, 1996*, Society of Fire Protection Engineers, Bethesda, Maryland, 1998.

Savornin, J., et al., "Analysis of Fire Risk in French Pressurized Water Reactors," *International Conference on Thermal Reactor*

Safety, Avignon, France, October 2–7, 1988, October 1988.

Simcox, S., N. Wilkes, and I. Jones, "Computer Simulation of the Flows of Hot Gases From the Fire at King's Cross Underground Station," pp. 19–25 in *Institution of Mechanical Engineers, King's Cross Underground Fire: Fire Dynamics and the Organization of Safety,* London, 1989.

Siu, N., and G. Apostolakis, "A Methodology for Analyzing the Detection and Suppression of Fires in Nuclear Power Plants," *Nuclear Science and Engineering,* Vol. 94, pp. 213–226, 1986.

Snell, J.E., V. Babrauskas, and A. Fowell, "Elements of a Framework for Fire Safety Engineering," pp. 447–456 in *Proceedings of Interflam 93, The Fourth International Fire Science and Engineering Conference,* Interscience Communications Limited, London, 1993.

Society of Fire Protection Engineers, *The SFPE Handbook of Fire Protection Engineering,* Second Edition, Vol. 3, pp. 201 and 203, Bethesda, Maryland, 1995.

Steckler, K.D., et al., "Fire Induced Flows Through Room Openings—Flow Coefficients," NBSIR 83-2801, National Bureau of Standards, Washington D.C., 1984.

Stroup, D.W., "Using Computer Fire Models To Evaluate Equivalent Levels of Fire and Life Safety," *Proceedings, Symposium of Computer Applications in Fire Protection Engineering,* Worcester Polytechnic Institute, Worcester, Massachusetts, June 1993.

Stroup, D.W., "Using Field Models to Simulate Enclosure Fires," *SFPE Handbook of Fire Protection Engineering,* Second Edition, National Fire Protection Association, Quincy, Massachusetts, pp. 3-152 to 3-159, 1995.

Sullivan, K., et al., "Inspection Techniques for Post-Fire Safe Shutdown of Nuclear Generating Stations," Technical Report A-3270-1, Brookhaven National Laboratory, Upton, New York, January 1989.

Traw, Jon S., "Future Perspective of the U.S. Model Building Codes," *Proceedings of the International Conference on Performance-Based Codes and Fire Safety Design Methods, Ottawa, Canada, September 1996,* Society of Fire Protection Engineers, Bethesda, Maryland, 1998.

____, "NFPA's Future in Performance-Based Codes and Standards—Report of the NFPA In-House Task Group," National Fire Protection Association, Quincy, Massachusetts, July 1995.

U.S. Atomic Energy Commission, WASH-1400 (now NUREG-75/014), "Reactor Safety Study: An Assessment of Accident Risks in U.S. Commercial Nuclear Power Plants," Washington, D.C., 1975.

U.S. Department of Commerce, "FASTLite," Special Publication 899, National Institute of Standards and Technology, Building and Fire Research Laboratory, Fire Modeling and Applications Group, Gaithersburg, Maryland, 1996.

U.S. Department of Defense, "Military Handbook, Reliability Prediction of Electronic Equipment," MIL-HDBK-217E, January 1990.

U.S. Department of Transportation, "Combustibility of Electrical Wire and Cable for Rail Rapid Transient Systems," DOT-TSC-UMTA-83-4-1, May 1983.

U.S. Nuclear Regulatory Commission, AEOD/S97-03, "Special Study, Fire Events—Feedback of U.S. Operating Experience," J.R. Houghton, June 1997.

———, Branch Technical Position APCSB 9.5-1, "Guidelines for Fire Protection for Nuclear Power Plants Docketed Prior to July 1, 1976."

———, Branch Technical Position CMEB 9.5-1, "Guidelines for Fire Protection for Nuclear Power Plants Docketed Prior to July 1, 1976," Appendix A.

———, "Elimination of Requirements Marginal to Safety—Solicitation of Public Comments,"

Federal Register, Vol. 57, p. 4166, February 4, 1992.

———, "Fire Protection Program for Operating Nuclear Power Plants," Final Rule 10 CFR Part 50. *Federal Register*, Vol. 45, p. 76602, November 19, 1980.

———, "Safety Goals for the Operation of Nuclear Power Plants: Policy Statement," *Federal Register*, Vol. 51, p. 28041, August 4, 1986.

———, Generic Letter 81-12, "Fire Protection Rule (45 FR 76602, November 19, 1980)," February 20, 1981.

———, Generic Letter 83-33, "NRC Positions on Certain Requirements of Appendix R to 10 CFR 50," October 19, 1983.

———, Generic Letter 85-01, "Fire Protection Policy Steering Committee Report," January 9, 1985.

———, Generic Letter 86-10, "Implementation of Fire Protection Requirements," April 24, 1986.

———, Generic Letter 88-12, "Removal of Fire Protection Requirements From Technical Specifications," August 2, 1988.

———, Generic Letter 88-20, "Individual Plant Examination for Severe Accident Vulnerabilities—10 CFR 50.54(f)," November 23, 1988.

———, Generic Letter 88-20, Supplement 4, "Individual Plant Examination of External Events (IPEEE) for Severe Accident Vulnerabilities—10 CFR 50.54(f)," June 28, 1991.

———, IE Information Notice 83-41, "Actuation of Fire Suppression System Causing Inoperability of Safety Related Equipment," June 22, 1983.

———, IE Information Notice 83-69, "Improperly Installed Fire Dampers at Nuclear Power Plants," October 21, 1983.

———, IE Information Notice 84-09, "Lessons Learned From NRC Inspections of Fire Protection Safe Shutdown Systems (10 CFR 50, Appendix R)," February 13, 1984.

———, Memorandum from L. Joseph Callan to Commissioners, "Preliminary IPEEE Insights Report," January 20, 1998.

———, NUREG-0050, "Recommendations Related to the Brown's Ferry Fire," February 1976.

———, NUREG-0675, Supplement 23, "Safety Evaluation Report Related to the Operation of Diablo Canyon Nuclear Power Plant, Units 1 and 2," June 1984.

———, NUREG-0781, Supplement 4, "Safety Evaluation Report Related to the Operation of South Texas Project, Units 1 and 2," July 1984.

———, NUREG-0800 (formerly NUREG-75/087), "Standard Review Plan for the Review of Safety Analysis Reports for Nuclear Power Plants—LWR Edition," July 1981.

———, NUREG-1150, Vol. 1, "Severe Accident Risks: An Assessment for Five U.S. Nuclear Power Plants," Vol. 1, December 1990.

———, NUREG-1493, "Performance-Based Containment Leak-Test Program," September 1995.

———, NUREG-1547/NISTIR 5842, "Methodology for Developing and Implementing Alternative Temperature-Time Curves for Testing the Fire Resistance of Barriers for Nuclear Power Plant Applications," L.Y. Cooper and K.D. Steckler, August 1996.

———, NUREG/CP-0129, "Proceedings of the Workshop on Program for Elimination of Requirements Marginal to Safety," M. Dey, F. Arsenault, M. Patterson, and M. Gaal, September 1993.

———, NUREG/CP-0138, "Proceedings of Workshop 1 in Advanced Topics in Risk and Reliability Analysis," October 1994.

————, NUREG/CR-2258, "Fire Risk Analysis for Nuclear Power Plants," M. Kazarians and G. Apostolakis, September 1981.

————, NUREG/CR-2300, "PRA Procedures Guide," January 1983.

————, NUREG/CR-3192, "Investigation of Twenty-Foot Separation Distance as a Fire Protection Method as Specified in 10 CFR 50, Appendix R," D.D. Cline, W.A. von Riesmann, and J.M. Chavez., 1983.

————, NUREG/CR-4230, "Probability-Based Evaluation of Selected Fire Protection Features in Nuclear Power Plants," M.A. Azarm and J.L. Boccio, May 1985.

————, NUREG/CR-4330, "Review of Light Water Reactor Regulatory Requirements: Identification of Regulatory Requirements That May Have Marginal Importance to Risk," Vol. 1, M.F. Mullen, W.B. Scott, W.B. Andrews, W.E. Bickford, A.J. Boegel, W.W. Little, P.J. Pelto, and T.B. Powers, 1986.

————, NUREG/CR-4550, "Analysis of Core Damage Frequency: Surry Power Station, Unit 1 External Events," Vol. 3, Rev. 1, Part 3, SAND86-2084, M.P. Bohn, J. A. Lambright, S. L. Daniel, et al., Sandia National Laboratories, December 1990.

————, NUREG/CR-4550, "Analysis of Core Damage Frequency: Peach Bottom Unit 2 Internal Events," Vol. 4, Rev. 1, Part 1, A.M. Kolaczkowski, W.R. Cramond, T.T. Sype, K.J. Maloney, T.A. Wheeler, and S.L. Daniel, Sandia National Laboratories, August 1989.

————, NUREG/CR-4550, "Analysis of Core Damage Frequency: Peach Bottom Unit 2 External Events," Vol. 4, Rev. 1, Part 3, SAND86-2084, M. P. Bohn, J.A. Lambright, S. L. Daniel, et al., Sandia National Laboratories, December 1990.

————, NUREG/CR-4598, "A User's Guide for the Top Event Matrix Analysis Code (TEMAC)," R.L. Iman and M.J. Shortencarier, August 1986.

————, NUREG/CR-4627, "Generic Cost Estimates: Abstracts From Generic Studies for Use in Preparing Regulatory Impact Analyses," Rev. 1, E. Claiborne, F. Sciacca, G. Simon, and G. Baca, February 1989.

————, NUREG/CR-4679, "Quantitative Data on the Fire Behavior of Combustible Materials Found in Nuclear Power Plants: A Literature Review," S.P. Nowlen, February 1987.

————, NUREG/CR-4681, "Enclosure Environment Characterization Testing for Base Line Validation of Computer Fire Simulation Codes," S.P. Nowlen, 1987

————, NUREG/CR-4832, "Analysis of the LaSalle Unit 2 Nuclear Power Plant: Risk Methods Integration and Evaluation Program (RMIEP), Summary," Vol. 1, A.C. Payne, Jr., July 1992.

————, NUREG/CR-4832, "Analysis of the LaSalle Unit 2 Nuclear Power Plant: Risk Methods Integration and Evaluation Program (RMIEP), Internal Fire Analysis," Vol. 9, J.A. Lambright, D.A. Brosseau, A.C. Payne, Jr., and S.L. Daniel, March 1993.

————, NUREG/CR-5088, "Fire Risk Scoping Study: Investigation of Nuclear Power Plant Fire Risk, Including Previously Unaddressed Issues," J.A. Lambright, S.P. Nowlen, V.F. Nicolette, and M.P. Bohn, Sandia National Laboratories, January 1989.

————, NUREG/CR-5384, "A Summary of Nuclear Power Plant Fire Safety Research at Sandia National Laboratories, 1975–1987," S.P. Nowlen, 1989.

————, NUREG/CR-5775, "Quantitative Evaluation of Surveillance Test Intervals Including Test-Caused Risks," I.S. Kim, S. Martorell, W.E. Vesely, and P.K. Samanta, November 1991.

————, NUREG/CR-5813, "Integrated Reliability and Risk Analysis System (IRRAS) Version 4.0: Reference Manual," Vol. 1, K.D.

Russell, M.K. McKay, M.B. Sattison, N.L. Skinner, S.T. Wood, and D.M. Rasmuson, January 1992.

———, Regulatory Review Group - Final Report, Volume Two, "Regulations," F.P. Gillespie et al., August 1993.

———, SECY 83-269, "Fire Protection Rule for Future Plants," July 5, 1983.

———, SECY 93-028, "Institutionalization of Requirements Marginal to Safety," February 5, 1993.

———, SECY 93-143, "NRC Staff Actions to Address the Recommendations in the Report on the Reassment of the NRC Fire Protection Program," May 21, 1993.

———, SECY 93-167, "Regulatory Analysis Guidelines of the U.S. Nuclear Regulatory Commission," June 14, 1993.

———, SECY 94-090, "Institutionalization of Continuing Program for Regulatory Improvement," March 31, 1994.

———, Translation 3383, "Evaluation of the COMPBRN Code, Analysis of the Performance of the Code" V. Cervantes, Bertin & Company, France, 1993.

———, Translation 3384, "Evaluation of the COMPBRN Code, Analysis of the Modeling File," V. Cervantes, Bertin & Co., France, 1993.

United Kingdom, "Housing and Building Control Act of 1984," Statutory Instrument (SI) 1985, No. 1965, Her Majesty's Stationery Office, London, 1985.

Wakamatsu, T., "Development of Design System for Building Fire Safety," pp. 881–898 in *Proceedings of the Second International Symposium on Fire Safety Science,* Hemisphere Publishing Corp., New York, 1989.

Walton, G., "CONTAM 93 User Manual," NISTIR 5385, National Institute of Standards and Technology, Gaithersburg, Maryland, March 1994.

Walton, W.D., K.B. McGrattan, and J.V. Mullin, "ALOFT-PC: A Smoke Plume Trajectory Model for Personal Computer," pp. 987-997 in *Proceedings, 19th Arctic and Marine Oilspill Program (AMOP) Technical Seminar,* Vol. 2, Environment Canada, Ottawa, Ontario, 1996.

Zukowski, E.E., T. Kubota, and B. Cetegen, "Entrainment in Fire Plumes," *Fire Safety Journal,* Vol. 3, pp. 107–121, 1980/1981.

Appendix A

REVIEW OF FIRE PROTECTION LITERATURE

CONTENTS

Figures

APPENDIX A
REVIEW OF FIRE PROTECTION LITERATURE

A.1 INTRODUCTION

A number of countries are developing, or already have adopted, performance-based fire codes. One of the benefits is designs to achieve fire safety that are better or less expensive than prescriptive codes. Generally the goal is "equivalency" with the prescriptive code, although it is realized that, in most cases, the effectiveness of the existing code is not known. Where possible, designers using performance-based methods are instead basing designs on qualitative "objectives" and quantitative "requirements." Expertise to confirm that these goals have been met exceeds the qualifications of people involved in traditional code enforcement. The Japanese Ministry of Construction, in the forefront of this effort, uses panels of experts and local officials to review performance-based designs submitted for approval. In New Zealand, an aggressive effort is under way to enhance what code officials know.

Performance-based design is now feasible because of the state of the art of fire prediction calculations, probabilistic risk assessment (PRA) techniques, and plant experience for ignition and suppression probabilities. This is illustrated in one reference (Bateman et al., 1993), which utilizes many of the techniques required for performance-based design to update PRAs of two existing nuclear power plants.

Most of the references chosen for this review were published between 1989 and 1998, illustrating the modern surge of interest in and capability of performance-based fire safety design.

Many countries, especially the United States, Japan, Australia, Canada, New Zealand, the United Kingdom, Sweden, and Finland, are developing detailed methodologies which could be used to evaluate the safety (and thus the code equivalency) of innovative building designs (SFPE, 1998a & 1998b). These methodologies will initially supplement the existing codes and will be used in innovative construction projects.

Success will lead eventually to performance-based codes for general use. The performance-based New Zealand code is already the only official code in that country. The Japanese effort is a major project for the Ministry of Construction and the Building Research Institute. Details of the performance-based methodology are being finalized during the design of major Japanese governmental facilities.

If good data on fire losses exist, the performance codes are tested against those data. If the data do not exist, the calculations are tested against calculated fire safety in buildings built to the existing codes, which are assumed to provide an acceptable degree of fire safety. As a result, the necessity of quantifying fire safety with such sensitive concepts as the value of human life is precluded.

Performance codes require that the fire safety design be tested against a set of criteria and scenarios which depend on the occupancy class of the structure. So there are differences in the rooms, ventilation, ignition sources, and framework for analysis, as well as differences in the criteria for success. On the other hand, there are many similarities, such as the mathematical models available for use, fire growth curves, and the concepts of hazard and risk. In this appendix, the methodologies being developed for performance-based regulation of residential and commercial occupancies in various countries are presented.

A.2 INTERNATIONAL PROGRAMS FOR RESIDENTIAL AND COMMERCIAL FIRE CODES

Each country discussed below has a single national fire code and an organization to maintain it, and has initiated the process of utilizing performance-based design methods.

New Zealand

Buchanan (1993) describes a new performance-based code introduced in New Zealand. The code requires specific fire engineering design for certain buildings and permits it as an option for all buildings. As with other performance codes, the New Zealand code was designed to

(1)   state its objectives clearly
(2)   specify performance requirements
(3)   permit any solution that meets the performance requirements

There is an important tradeoff between accuracy and simplicity in the design process. A complicated code may give an illusion of accuracy that cannot be achieved. The New Zealand code is a major step in the right direction; it has excellent structure but does not specify quantification of performance or safety.

The 1991 New Zealand Building Act is concerned with the health and safety of building occupants, covering structural stability, access, user safety, services, and facilities. Secondary concerns are energy efficiency, fire fighting access, and the prevention of fire spread to other buildings. There are no controls on fire spread or damage within the fire building.

The code uses a five-level structure:

(1)   objective
(2)   functional requirements
(3)   performance
(4)   verification method
(5)   acceptable solution

The first three are mandatory; the last two can reference existing standards. Each fire must be analyzed in four categories:

(1)   outbreak of fire
(2)   means of escape
(3)   spread of fire
(4)   structural stability during fire

A design guide to provide guidance to those making or reviewing specific designs to meet the code is being produced by a "study group." The overall strategy of the design guide is shown in Figure A.1. Given the subjectivity involved in the designs, a need has been identified to achieve consistency in the design and approval process (Caldwell, 1998).

The ability of a design to continue to satisfy, given changes in the use of the structure, is known as "durability." Recognizing this problem, the New Zealand code places a 10-year limit on the legal liability of the designer (Hunt, 1998). Insurance companies are developing 10-year insurance plans.

A substantial educational effort is being implemented in New Zealand comprising a periodic 5-day workshop and the establishment of a 1-year master of engineering degree at the University of Canterbury, the latter for those who already have a relevant bachelors engineering degree.

Australia and Canada

These two countries are discussed together because of the close coordination between their professional staffs in the development of performance code methodology. Beck (1991) describes the joint effort. This initiative is more complicated than the New Zealand method, and has not yet been implemented. The method utilizes as a framework a central risk assessment model (FIRECAM) that evaluates quantitative information from six submodels as shown in Figure A.2.

A level of redundancy is required so as not to rely solely on a single component or subsystem, but too much redundancy would be too costly. In a recent paper, Thomas and Bowen (1998) comment that since "performance" is sometimes impossible to quantify, the code should be known as an "objective-based" code. Canada plans to publish "intent" statements in 1998, and an objective-based code in 2001 (Chauhan, 1998). A limitation is the maturity of technology, including calculation of fire growth, flame spread, combustibility of materials, and the use of models. Then it takes time to incorporate current knowledge into design. Simple equations are likely to be adequate in many cases, rather than

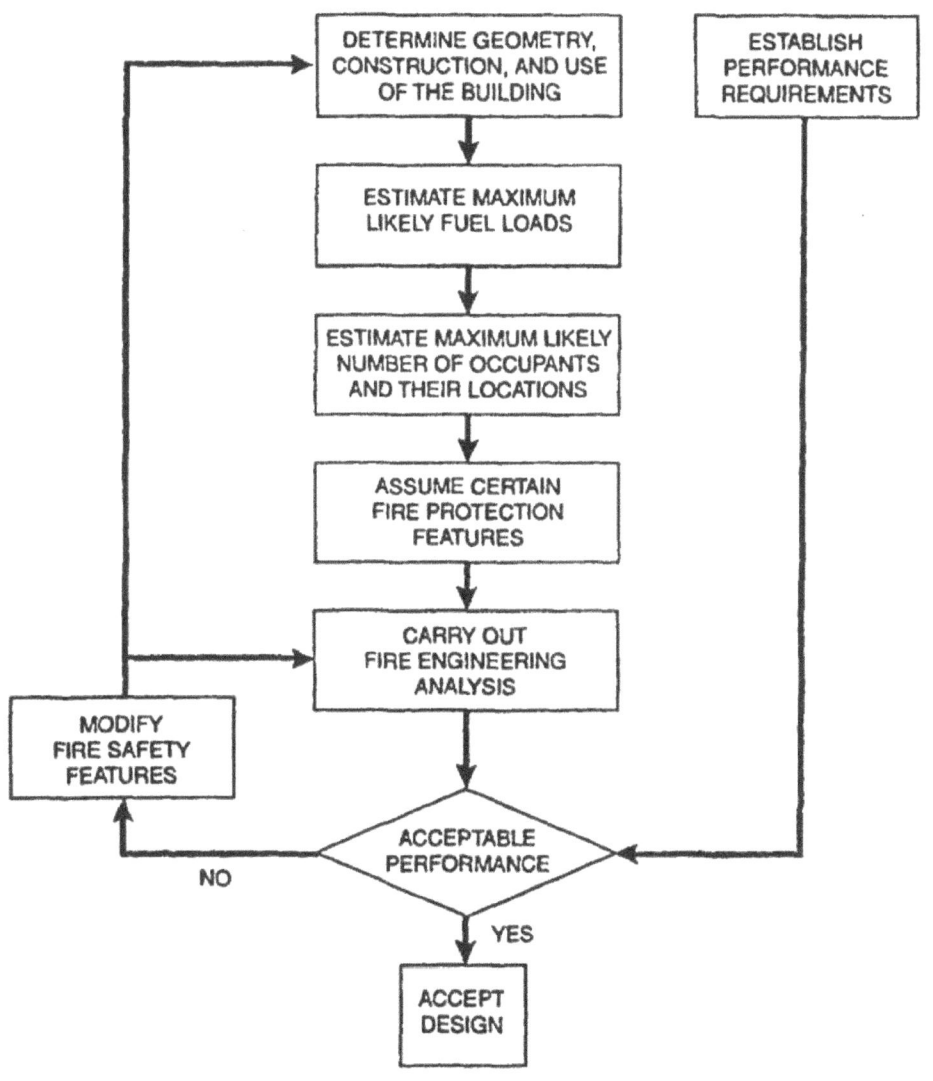

(Source: Buchanan, 1993. Reproduced by permission of author)

Figure A.1
Outline of Fire Engineering Design Procedure

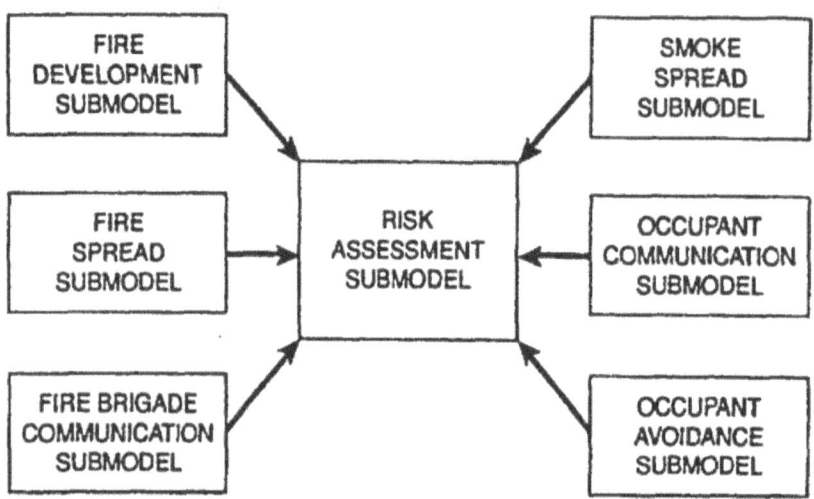

Source: Beck, 1991; reproduced by permission of author.

Figure A.2
A Risk Assessment Submodel

using complete models. The central FIRECAM uses event tree formulation; the timing from one event to another is supplied by the submodels. Two parameters are used: the "expected risk to life" (ERL), that is, fatalities over the expected life of the building, and "fire-cost expectation" (FCE), that is, the aggregate of all costs over the life of the building. ERL must be at least as good as the value from the prescriptive code, and FCE must be less.

Various available calculational codes are used in the submodels as applicable, but the authors have in some cases developed their own to increase speed and reduce calculating costs. The parameters calculated and computer codes used in each of the submodels are described in the reference. If a calculational model or data is not available, expert opinion is used. To date, calculations give estimates of risk to life safety that are significantly higher (worse) than values obtained by analysis of historical records.

Two papers on the Canadian method are presented in Interflam 93, the Sixth International Fire Conference. Cornelissen et al. (1993) consider four ways to look at the code equivalency of three-story nonresidential, wood frame buildings.

The buildings have 45-minute, 25-minute, 10-minute, or 10-minute endurance with sprinklers; 45 minutes is the code requirement, but 10 minutes with sprinklers turns out to be much better. The four ways are

(1)   The Building Fire Safety Model (R.W. Fitzgerald, Worcester Polytechnic Institute)

(2)   Building Code Assessment Framework (M. Katzin et al., ASTM STP 1150 (1990) pp. 234–237)

(3)   NIST Fire Risk Assessment Model (HAZARD I)

(4)   NRCC Fire Risk-Cost Performance Assessment Model (National Research Council of Canada)

Number 4 was chosen because it minimizes the use of subjective input data. It is a systems approach using fire dynamics, building design, active and passive fire safety features (and the cost of maintaining the active ones), and human behavior. Building codes are used as the reference level of safety.

Yung et al. (1993) also describe the use of the National Research Council of Canada (NRCC) Fire Risk-Cost Performance Assessment Model to evaluate various fires with door open/closed and people awake/asleep, and compares various code-compliant designs. Cost was not evaluated in this study. Using the NRCC method, Yung determined that a 1-hour wood frame building with a central alarm system connected to the fire department, with or without sprinklers, is better than masonry without the active features.

The NRCC model uses statistical data on fire starts and time of start (awake/asleep). The model then calculates fire growth, smoke spread, and detector and sprinkler actuation. The fire is truncated when the sprinkler actuates or the fire brigade arrives. The egress time available is calculated as hazard time minus alarm time. Yung considered six design fire (smoldering, flaming, flashover) times (door open, door closed). When sprinklers activated, some of the flashover and flaming fires were rendered non-lethal. The worst situation is when the fire flashes over and the occupants are asleep.

The following models are used in the NRCC Fire Risk-Cost Performance Assessment Model:

(1) smoke movement model, which calculates time to untenable conditions at various places

(2) fire detection model, which calculates smoke detector and sprinkler activation time and flashover time

(3) occupant warning and response model, with probability of warning from model 2

(4) fire brigade action model

(5) smoke hazard model

(6) evacuation duration model, which calculates the time to get all the people out

(7) egress model

(8) boundary element model (wall endurance between compartments)

(9) fire spread model based on model 8 and the fire brigade to estimate property loss

Grubits (1993) reports comprehensive, specific plans for performance-based fire regulation reform in Australia. Although 70 percent of the Australian building code is fire related and it was recognized in 1989 that fire regulation reform was needed, lack of funding prevented change. It is expected that $0.5 million (or more) (Australian dollars) per year will now be made available for the development of new, more flexible regulatory provisions. A 1-percent savings in building costs is expected, corresponding to savings of $370 million (Australian dollars) per year. A risk assessment methodology will be used, with acceptable values of risk to life obtained by evaluating reference buildings. New material test methods will be specified in which the data can be used in performance calculations. Preference will be given to the latest generation of International Standards Organization (ISO) tests such as use of the cone calorimeter.

In a recent paper, Allen, Grubits, and Quaglia (1998) comment that experience shows that the regulatory authorities need to participate in the evolution of performance-based codes, and that the cost savings in using such codes averages 5–6 percent of the cost of the building. In one building, in Brisbane, the saving was 8 percent. Reports on the progress on experience with Australia's performance-based regulations and the development of the supporting fire engineering guidelines have been recently provided (Graham, 1998 and Johnson, 1998).

Japan

According to Bukowski (1993), the Japanese system is described in four volumes published in 1988 (in Japanese). The volumes consider

- smoke control and evacuation safety
- prevention of the outbreak and development of fire

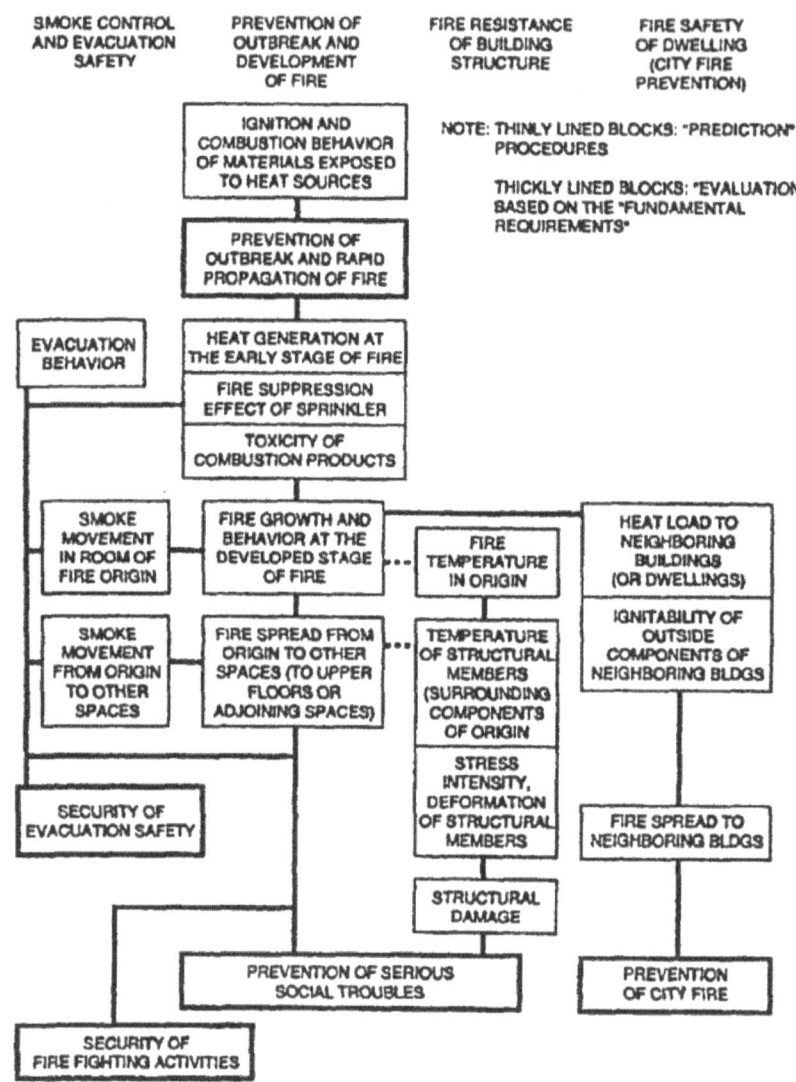

Source: Bukowski, 1993. Reproduced by permission of author.

Figure A.3
The Japanese Evaluation Procedure

- fire resistance of the building structure
- prevention of fire spread to other buildings

A schematic diagram of the structure of the Japanese evaluation procedure is shown in Figure A.3. Wakamatsu (1989) describes the Japanese effort in detail.

The Ministry of Construction organized a 5-year research program in 1982 to develop a performance design system. The effort included professional staff from the Ministry, the Building Research Institute (BRI), and two nonprofit organizations. More than 100 experts on fire research and engineering, architects, and people in related professions served on the committees.

The purpose of this program is to develop a national evaluation method for fire safety in buildings, rather than further improvement in fire safety. The Japan Building Standard Law is the appropriate level of fire safety. Specifications for the performance methods were decoded from specific articles of this law. The objectives of the law are

(1)   prevention of fire outbreak
(2)   human safety in fire
(3)   prevention of "public troubles"
(4)   prevention of property losses

An example of "public troubles" would be burning down a neighborhood or interfering with another tenant in the building.

Provisions for fire fighting and fire brigade accessibility to a building are also taken as basic requirements. Fire fighting is required and is also expected to work as a "trump" when other measures do not control a fire.

The framework used to organize and document the approach and solution for each problem area of fire protection engineering basically comprises the following:

(1)   fundamental requirements
(2)   technical standards for engineering evaluation
(3)   prediction method of relevant fire phenomena
(4)   concepts of testing methods

These, allowing for translation, are parallel to the steps developed by Gross (1975) at NIST for performance-based regulation in building construction.

Subjects of the proposed predictive methods (number 3 above) are in the list that follows: "Subjects of Predictive Methods for Fire Safety Design." The list, taken from Wakamatsu (1989), names the fire, smoke, and structural effects that should be considered in a complete fire safety performance evaluation of a major occupied structure. NRC concerns for fire safety in nuclear power plants would encompass only selected subjects from this table.

Because of the wide range of expertise needed to deal with the broad range of fire safety concerns listed, the Japanese organized five separate

committees to develop the methodologies needed. Their responsibilities are detailed and the design procedure trees of four of them are presented in the Wakamatsu paper.

The methodology has actually been applied (as it has evolved) to major structures planned in Japan. One of these is a "National Theater" (56,000 m² (603,00 ft²) with three auditoriums, stores, and other features similar to the Kennedy Center in Washington, D.C.). The Japanese found they could use wood lining in the ballet and opera theaters, which is not allowed under the prescriptive Building Standard Law.

Performance calculations are allowed under an equivalency clause in the Japanese National Fire Code, but local code officials can have difficulty deciding whether a performance design is indeed "equivalent." When these questions arise, the Ministry of Construction is consulted. The Ministry assembles a panel, consisting of people from its own staff, the Building Research Institute (BRI), universities, and the affected local officials, to review the performance calculation and approve or disapprove.

Bukowski and Brabauskas (1994) presents, as appendices, translations of the tables of contents of the four volumes of the Japanese report. The technical summary was translated into English by its author, a prominent Japanese modeler, Dr. Takeyoshi Tanaka (1989). Article 38 of the Japanese Construction Code allows "equivalency" in designing safety features, so the four-volume set is known as the "equivalency" report. Tanaka and Harada (1998) have participated in an international "case study" in which the participants designed (using three different methods) a four-story office building with an atrium: (1) according to their prescriptive code, (2) according to a performance method, with detectors and no sprinklers, and (3) with sprinklers and no detectors. The work required 375 person-days, 225 for design and 150 for verification to Japanese standards. They did not use a complete computer model, but used simple mathematical correlations because the architects on the team were not comfortable with the computer models.

Subjects of Predictive Methods for Fire Safety Design

1. Combustion and fire behaviors
    1.1 Ignition of combustible materials
    1.2 Behavior of turbulent diffusion flame
        1: Flame height
        2: Temperature and velocity on axis of turbulent diffusion flame and fire plume
        3: Amount of smoke included in turbulent diffusion flame and buoyant flow
        4: Radiative energy from turbulent diffusion flame
    1.3 Formation of smoke layer and ventilation
    1.4 Heat transmission at early stage of fire
        1: Spread of burning area at early stage of fire
        2: Radiative heat transfer to surroundings
        3: Convective heat transfer to ceiling, wall, floor, and other surfaces exposed to fire
    1.5 Flame spread
        1: Velocity of upward spread of flame on vertical surface
        2: Velocity of steady-state spread of flame on surface with arbitrary heat flux distribution
        3: Velocity of steady-state spread of flame on surface receiving constant heat flux
    1.6 Effectiveness of automatic sprinkler
        1: Response time of fire extinction equipment
        2: Time required for fire suppression
        3: Properties of fire extinguishing equipment
    1.7 Burning behavior at developed stage of fire
        1: Standard fire temperature prescribed in the enforcement order of the Building Standard Law
        2: Models by Ingberg, Kawagoe, Magnusson, Babrauskas, Harmathy, etc.
    1.8 Fire spread between buildings
        1: Fire spread due to radiative heat transfer
        2: Standardization of heat condition
        3: Behavior of external flames
        4: Behavior of flame rising up from a burning structure
2. Smoke movement and smoke control
    2.1 Single layer models
        1: Steady-state model for multiple compartments on multiple floors
        2: Unsteady-state model for multiple compartments on multiple floors
    2.2 Two layer model
        1: Unsteady-state model for multiple compartments on single floor
    2.3 Simplified model for evaluating smoke control systems
3. Evacuation behavior
    3.1 Model of evacuee (properties, distribution, velocity of evacuees)
    3.2 Model of evacuation spaces or routes (room, path, stairs, hall, vestibule, lines, and crowding)
    3.3 Model of evacuation behavior
        1: Start time of evacuation
        2: Evacuees' movement in a unit space
4. Fire resistance of building structure
    4.1 Fire temperature as heat load to the structure (which is given on the basis on the line 1.7 "burning behavior at developed stage of fire")
    4.2 Temperature of structural members
        1: Model for reinforced-concrete members (one/two dimensional heat flow)
        2: Model for steel structural members
        3: Model for assembled structural members (for example, a structure assembled by reinforced-concrete slab and steel beams)
    4.3 Thermal stress and deformation
        1: Model for reinforced-concrete members
        2: Model for steel structural members
        3: Model for assembled structural members
5. Fire safety performance of dwellings
    5.1 Evaluation safety performance for evacuation safety in dwellings
        1: Evaluation of difficulty for evacuation
    5.2 Evaluation model for performance of fire prevention

Source: Wakamatsu, 1989. List reproduced with permission of author.

In the performance-based designs they found they could remove limitations on the size of the fire compartment, and some walls could be lighter than standard. They needed to increase the height of the atrium and provide water curtains for the glass atrium walls. They also found that, in a few instances, the prescriptive design was inadequate. However, they felt that for this ordinary building, the effort for the performance-based design was not worthwhile.

The Japanese Building Research Institute has an ongoing program on fire testing methods for materials and structures that will meet the performance criteria (Nakaya, 1993). The new program will also make contributions to the international "harmonization" of testing and assessment procedures, so that the same products and procedures will be acceptable in many countries. These will be comprehensive and may help industry to develop new types of products and designs and allow engineers increased freedom in fire safety design.

Nordic Countries

As can be seen from the list of references at the end of this appendix, a number of papers come from a symposium held at Espoo, Finland, in August 1993. The Nordic countries are cooperating in the development of a performance-based fire code that will eventually be adopted by each country. The technique for design and for regulation is much less formal than that used by the Japanese. The designers, any consultants, and the local regulators work together from the inception of a project. The group decides on the goals and requirements, the computer programs, and other methodology to be used to solve each problem. The group continues to work together as the project proceeds. The computer program most frequently used is a multiroom zone model developed in Japan, BRI-2.

S.E. Magnusson (1993) prepared one of the most comprehensive papers combining classical probability analysis with fire modeling. The introduction to his paper discusses concepts, lists ISO documents circa 1985, and concludes that, although progress in performance quantification is being made, the ISO documents are not state of the art.

S.E. Magnusson states, "Fire risk calculation comprises a wide range of deterministic and probabilistic methods; Chapter 4 of the SFPE [Structural Fire Protection Engineering] Handbook gives an excellent review. In this paper we will describe only two possible approaches: the first a demand-supply, reliability-based methodology originally developed for structural engineering design, the second an event tree logic extensively used in chemical industry quantitative risk analysis (QRA)."

This statement is followed by a very terse presentation of reliability theory and QRA-based design. Parameter uncertainty analysis is described as follows:

(1)   List all parameters that are potentially important contributors to uncertainty in model prediction.

(2)   Specify the maximum range of each parameter.

(3)   Subjectively adjust a probability distribution to the maximum range.

(4)   Derive quantitative statements about the effect of parameter uncertainty on model prediction.

(5)   Rank the parameters with respect to their contribution to the uncertainty in model prediction.

Steps 1–3 require an expert with a complete understanding of the model and the underlying database. In Magnusson's opinion, prospects for applying fire safety engineering principles seem good on the component or subsystem level, and continued rapid development is expected. Problems will arise when discussing to what extent accepted performance of all involved subsystems amounts to acceptance of the whole building. He feels that a performance-based code at the whole-building level is probably more than 5 years away. Much more work needs to be done in the following areas:

- completeness of analysis (identification of all significant event sequences)
- treatment of uncertainty
- relation between prescription and performance-based parts of the code

He lists a number of items for which preparatory work is being done for international pre-standardization and standardization.

Magnusson (1998) continues to pioneer in the field of developing means to convert "hazard" to "risk." Recently, Frantzich (1998), also from Lund University in Sweden, reported on this work. Currently they are deriving safety factors (or uncertainty factors) for risk assessment by a method called "First Order–Second Moment" analysis. An n-dimensional "failure surface" by a Monte Carlo series of calculations is created, and indices for the relevant variables are obtained. These are used to derive safety factors.

International Coordination

International coordination activities on performance-based fire safety design are proceeding under two organizations, the International Council for Building Research and Development (CIB) and the International Standards Organization (ISO). CIB has created four subgroups under Committee W14 to provide a strategic overview of fire technology needs over the next 10 years (Kokala, 1998). W14 has more than 50 members from 30 countries, and organizes workshops open to all interested persons.

The following are the four subgroups under Committee W14:

(1) Engineering Evaluation of Performance-Based Systems—Chm.: R. Bukowsky, USA

(2) Verification of Computer Codes for Predicting Fire Development and Smoke Movement—Chm.: Keski-Rahkonen, Finland

(3) Thermal Response of Structures—Chm.: Wickstrom, Sweden

(4) Laboratory Calibrations & Measurements—Chm.: Hasemi, Japan

The following two new subgroups may be formed:

- Quantification of Uncertainty—Chm.: Magnusson, Sweden
- Codes for Fire Resistance in Buildings—Chm.: Kruppa, France.

W14 is carrying out a round-robin, currently on simple problems, to compare the results with 13 different fire models. Early results vary by a factor of 2.

ISO Technical Committee 92, Subcommittee 4 (ISO/TC92/SC4), "Fire Safety Engineering," has the goal of developing reports containing the framework for cost-effective, safe, enviromentally benign, fire safety design. ISO does no research; instead its committees are weighted toward the regulators, fire brigades, building designers, constructors, building managers, and insurers (Becker, 1998). ISO will also evaluate and validate computational models.

A.3 CONCLUSIONS

There is effort worldwide in utilizing the existing capability to predict fire and smoke spread and to calculate the resulting hazards in performance-based fire codes. Generally, the degree of safety desired is based on equivalency with the existing prescriptive codes, although it is recognized that, in some cases, improved safety could be attained. Most developed countries, other than the United States, have national fire codes and governmental organizations to administer them, simplifying the accommodation of political and policy changes. The Japanese Ministry of Construction, with help from the Japanese Building Research Institute and universities, has extensive efforts.

Experience indicates that it is more difficult to develop and regulate performance-based designs than to use prescriptive codes. The performance-based process requires more technical expertise and analyses. The qualitative requirements, the quantitative criteria to meet these requirements, and finally, the method of evaluating whether the criteria have been met must be developed. In general, one must examine the effects of a number of fires of the largest feasible size in each of several feasible locations to see if the selected fire protection provisions will provide safety. In order to verify the design, records or commentary is needed at each step. Most authors feel they can calculate the hazard created by a design, but not the risk. Because of uncertainties, factors of safety should be applied to the results of the design. These are related to risk.

A.4 BIBLIOGRAPHY

Allen, H., S. Grubits, and C. Quaglia, "Reflections on Ten Years of Fire Safety Engineering in Australia," *Proceedings of the International Conference on Performance-Based Codes and Fire Safety Design Methods, Ottawa, Canada, September 1996,* Society of Fire Protection Engineers, Bethesda, Maryland, 1998.

Bateman, K., W. Parkinson, S. Oh, and J. Haugh, "The Impact of Updated Fire Events Data and Modeling Techniques in the Electric Power Research Institute Fire PRA Method on Two Nuclear Power Plant Fire Risk Studies," Science Applications International Corp., Los Altos, California, 1993.

Beck, V.R., "Fire Safety System Design Using Risk Assessment Models: Developments in Australia,." International Association for Fire Safety Science, pp. 46–59 in *Proceedings of the 3rd International Symposium* (G. Cox and B. Langford, eds.), Elsevier Applied Science, London and New York, 1991.

Becker, W., "ISO/TC92/SC4 Fire Safety Engineering-Present Activities and Future Strategy," *Proceedings of the International Conference on Performance-Based Codes and Fire Safety Design Methods, Ottawa, Canada, September 1996,* Society of Fire Protection Engineers, Bethesda, Maryland, 1998.

Buchanan, A. H., "Fire Engineering for a Performance-Based Code," *Proceedings of the Sixth International Conference on Fire Safety,* Interscience Communications Limited, London, 1993.

Bukowski, R., "A Review of International Fire Risk Prediction Methods," *Proceedings of the Sixth International Conference on Fire Safety,* Interscience Communications Limited, London, 1993.

Bukowski, R., and V. Babrauskas, "Developing Rational Performance-Based Fire Safety Requirements in Model Building Codes," *Fire and Materials* , Vol. 18, No. 3, pp. 173–191, May/June 1994.

Bukowski, R., and T. Tanaka, "Toward the Goal of a Performance Fire Code," *Fire and Materials*, Vol. 15, No. 4, pp. 175–180, October–December 1991.

Bukowski, R.W., "Setting Performance Code Objectives—How Do We Decide What Performance the Codes Intend," p. 555 in *Proceedings of Interflam 96, The Seventh International Fire Science and Engineering Conference*, Cambridge, England, March 1996.

Caldwell, C. A., "Fire Engineering Performance Based Design Guidelines for Design Submittals and Reviews," *Proceedings of the Second International Conference on Performance-Based Codes and Fire Safety Design Methods, Maui, Hawaii, May 1998*, Society of Fire Protection Engineers, Bethesda, Maryland, 1998.

Chauhan, R. B., "The Canadian Approach to Fire Safety - Objective-Based Building Codes for 2001," *Proceedings of the Second International Conference on Performance-Based Codes and Fire Safety Design Methods, Maui, Hawaii, May 1998*, Society of Fire Protection Engineers, Bethesda, Maryland, 1998.

Cornelissen, A.A., G.V. Hadjisophocleos, and D. Yung, "Risk-Cost Assessment for Non-Residential Buildings," pp. 427–435 in *Proceedings of Interflam 93, The Fourth International Fire Science and Engineering Conference*, Interscience Communications Limited, London, 1993.

Frantzich, H., "Fire Safety Risk Analysis of a Hotel; How to Consider Parameter Uncertainty," *Proceedings of the Second International Conference on Performance-Based Codes and Fire Safety Design Methods, Maui, Hawaii, May*

1998, Society of Fire Protection Engineers, Bethesda, Maryland, 1998.

Graham, R. T., G. C. Ramsay, and M. B. N. Horasan, "Australia's Performance-based Regulations - A Fire Safety Engineering Experience, *Proceedings of the Second International Conference on Performance-Based Codes and Fire Safety Design Methods, Maui, Hawaii, May 1998*, Bethesda, Maryland, 1998.

Gross, J.G., et al., "Interim Performance Criteria for Solar Heating and Combined Heating/Cooling Systems and Dwellings," prepared for Department of Housing and Urban Development by National Bureau of Standards, U.S. Government Printing Office, Washington, D.C., January 1975.

Grubits, S.J., "Fire Regulation Reform in Australia," *Nordic Fire Safety Engineering Symposium*, Espoo, Finland, September 1993.

Hunt, J. H., "Performance-Based Codes: The New Zealand Experience," *Proceedings of the International Conference on Performance-Based Codes and Fire Safety Design Methods, Ottawa, Canada, September 1996*, Society of Fire Protection Engineers, Bethesda, Maryland, 1998.

Johnson, P., T. Lovell, and I. Moore, "Performance-Based Fire Safety Design - Development of the "Fire Engineering Guidelines," *Proceedings of the Second International Conference on Performance-Based Codes and Fire Safety Design Methods, Maui, Hawaii, May 1998*, Bethesda, Maryland, 1998.

Katzin, M., et al., "Fire Hazard and Fire Risk Assessment," STP 1150, American Society for Testing and Materials, Philadelphia, Pennsylvania, pp. 234–237, 1990.

Kokkala, M., "CIB W14 Activities to Promote Performance-Based Fire Safety Design," *Proceedings of the International Conference on Performance-Based Codes and Fire Safety Design Methods, Ottawa, Canada, September 1996*, Society of Fire Protection Engineers, Bethesda, Maryland, 1998.

Magnusson, S.E., "Performance-Based Codes," pp. 413–425 in *Proceedings of Interflam 93, The Fourth International Fire Science and Engineering Conference*, Interscience Communications Limited, London, 1993.

Magnusson, S. E., "How to Derive Safety Factors," *Proceedings of the International Conference on Performance-Based Codes and Fire Safety Design Methods, Ottawa, Canada, September 1996*, Society of Fire Protection Engineers, Bethesda, Maryland, 1998.

Meacham, B. J., "Performance-Based Codes and fire Safety Engineering Methods: Perspectives and Projects of the Society of Fire Protection Engineers," *Proceedings of the International Conference on Performance-Based Codes and Fire Safety Design Methods, Ottawa, Canada, September 1996*, Society of Fire Protection Engineers, Bethesda, Maryland, 1998.

Nakaya, I., "Our Activities Toward Performance-Based Fire Regulation in Japan," *Proceedings of the Nordic Fire Safety Engineering Symposium*, Espoo, Finland, 1993.

National Fire Protection Association, "HAZARD I," Quincy Massachusetts.

National Fire Protection Association, *Life Safety Code*, Quincy, Massachusetts.

National Institute of Standards and Technology, "Interim Performance Criteria for Solar Heating and Combined Heating/Cooling Systems and Dwellings," U.S. Government Printing Office, Washington D.C., January 1, 1975.

Society of Fire Protection Engineers (SFPE), *Proceedings of the International Conference on Performance-Based Codes and Fire Safety Design Methods, Ottawa, Canada, September 1996*, SFPE, Bethesda, Maryland, 1998a.

Society of Fire Protection Engineers, *Proceedings of the Second International Conference on Performance-Based Codes and Fire Safety Design Methods, Maui, Hawaii, May 1998*, SFPE, Bethesda, Maryland, 1998b.

Tanaka, T., "A Performance-Based Design for Fire Safety in Buildings," pp. 151–175 in *Fire Safety and Engineering: International Symposium Papers*, Warren Center for Advanced Engineering, University of Sydney, Australia, 1989.

Tanaka, T., and K. Harada, "A Case Study Using the Performance-Based design System in Japan," *Proceedings of the International Conference on Performance-Based Codes and Fire Safety Design Methods, Ottawa, Canada, September 1996*, Society of Fire Protection Engineers, Bethesda, Maryland, 1998.

Thomas, R., and R. Bowen, "Objective-Based Codes: The Canadian Direction," *Proceedings of the International Conference on Performance-Based Codes and Fire Safety Design Methods, Ottawa, Canada, September 1996*, Society of Fire Protection Engineers, Bethesda, Maryland, 1998.

Wakamatsu, T., "Development of Design System for Building Fire Safety," pp. 881–898 in *Proceedings of the Second International Symposium on Fire Safety Science*, Hemisphere Publishing Corp., New York, 1989.

Yung, D., G. Hadjisophocleos, and H. Takeda, "Comparative Risk Assessment of 3 Story Wood Frame and Masonry Construction Apartment Buildings," pp. 499–508 in *Proceedings of Interflam 93, The Fourth International Fire Science and Engineering Conference*, Interscience Communications Limited, London, 1993.

Appendix B

CONTRIBUTION OF FIRE TO FREQUENCY OF CORE DAMAGE
IN OPERATING NUCLEAR POWER PLANTS: A DATABASE

CONTENTS

Tables

APPENDIX B
CONTRIBUTION OF FIRE TO FREQUENCY OF CORE DAMAGE IN OPERATING NUCLEAR POWER PLANTS: A DATABASE

B.1 INTRODUCTION

This appendix describes a database of the contribution of fire to the frequency of core damage in operating nuclear power plants. Section B.2 presents the database with a summary review of fire's contribution to core-damage frequency (CDF). Section B.3 presents a detailed review of a boiling-water reactor (BWR) whose fire contribution is significant. Section B.4 presents a detailed review of a pressurized-water reactor (PWR) whose fire contribution is significant. Section B.5 contains a list of references.

B.2 SUMMARY REVIEW OF SITES WITH FIRE ANALYSES

The contribution that fire makes to annual CDF is summarized in Table B.1. From this table, it can readily be seen that fire makes an important contribution to CDF at some plants (Limerick and LaSalle Unit 2).

This study searched 48 sites (in several cases, two plants at the same site are grouped in an individual plant examination (IPE) or a probabilistic risk assessment (PRA)). Most of the IPEs do not contain a fire analysis, and the only external event analyzed is internal flooding. From the 48 sites searched, 12 have a fire analysis

In this section, the 12 sites that have fire analyses are reviewed. Each of the sections that follow contains the following information about the site:

- total annual CDF

- total fire frequency contribution to the total annual CDF

- percentage of the total fire frequency contribution to the total annual CDF

- the locations at which the impact of fire is more important

The sites that follow are presented in order of descending percentage of fire contribution to annual CDF.

Indian Point Unit 2

The total mean CDF for Indian Point Unit 2 is approximately 9.6E-5 per reactor-year. The calculated annual CDF due to fire is 6.5E-5, or about 68 percent of the total. The impact of fire is important in the *electrical tunnel, switchgear room,* and *cable spreading room.*

Limerick Unit 1

The total mean CDF for Limerick Unit 1 is 4.4E-5 per reactor-year for all initiators. The total annual contribution to core damage, from all fires in all zones, is 2.3E-5 or about 53 percent of the total CDF. All of the three most dominant contributors to CDF are fire-induced sequences. Fires in the *13-kV switchgear room,* the *safeguards access area,* the *control rod drive (CRD) hydraulic equipment area,* and the *general equipment area* contribute more than 80 percent of the fire-induced CDF.

LaSalle Unit 2

The total mean CDF for LaSalle Unit 2 is 6.77E-5 per reactor-year. The estimated annual contribution to CDF from all fires in all zones is 3.21E-5. Fires in the *control room, Division 2 essential switchgear room, Division 1 essential switchgear room, and auxiliary equipment room* contribute more than 93 percent of the fire-induced CDF. Fires and internal initiating events are of roughly comparable importance in determining the CDF. Fires contribute to about 47 percent of the total CDF. Six of the ten dominating sequences are fire-induced sequences.

Table B.1 Plant Core-Damage Frequency (CDF)

Plant	Total CDF (per RY)	Fire CDF (per RY)	Contribution of Fire to Total CDF	Reference
Indian Point 2*	9.6E-5	6.5E-5	68 percent	Indian Point 2 IPE (Consolidated Edison, 1992)
Limerick 1	4.4E-5	2.3E-5	53 percent	Limerick PRA (NUS, 1983)
LaSalle 2	6.77E-5	3.2E-5	47 percent	NUREG/CR-4832, Vol. 1
Big Rock Point (BRP)	9.75E-4	2.3E-4	24 percent	BRP PRA (Consumers Power Company, 1981)
Peach Bottom	1.1E-4	2.0E-5	18 percent	NUREG-1150, Vol. 1
Seabrook	2.3E-4	2.5E-5	11 percent	Seabrook PRA (Garrick et al., 1983)
Zion	6.7E-5	4.6E-6	7 percent	Zion PRA (Commonwealth Edison Co., 1981)
Surry	1.96E-4	1.1E-5	6 percent	NUREG-1150, Vol. 1
Oconee	2.5E-4	1.0E-5	4 percent	Oconee PRA (Nuclear Safety Analysis Center, 1984)
South Texas Project (STP)	4.4E-5	4.9E-7	1 percent	STP IPEEE (Cross et al., 1992)
Catawba 1 and 2	7.8E-5	3.4E-7	< 1 percent	Catawba IPEEE (Duke, 1992)
McGuire	7.4E-5	8.1E-8	< 1 percent	McGuire IPEEE (Duke, 1991)

*The Indian Point Unit 2 (IP2) does not contain external events analyses. The fire contribution was taken from a report prepared by EG&G (EGG-2660) in 1991. The data in that report were based on a report prepared in the 1980s, and the total CDF was calculated as the CDF from the IP2 IPE (3.13E-5) plus the fire contribution (6.5E-5). The percentage was calculated for this study by using these values.

Big Rock Point

The total mean CDF for Big Rock Point is 9.75E-4 per reactor-year. The estimated annual contribution to CDF from all fires in all zones is 2.3E-4, or about 24 percent of the total. The impact of fire is important in the *station power room* and *cable penetration area within the containment.*

Peach Bottom

Peach Bottom's total mean CDF is 1.1E-4 per reactor-year. The estimated annual contribution to CDF from all fires in all zones is 2.0E-5, or about 18 percent of the total CDF. The impact of fire is especially important in the *emergency switchgear rooms, control room*, and *cable-spreading room.*

Seabrook

The total mean CDF as calculated in the Seabrook PRA is 2.3E-4 per reactor-year. Fire contributes 2.5E-5, or 11 percent of the total CDF. The impact of fire is an important initiator in the *control room*, the *primary component cooling water pump area, turbine building*, and *cable spreading room.*

Zion

The total mean CDF as calculated in the Zion PRA is 6.7E-5 per reactor-year. This includes an annual contribution of 4.6E-6 attributable to fire. Fire sequences comprise approximately 7 percent of the total CDF. The impact of fire is important in the *auxiliary electrical equipment room* and the *inner and outer cable-spreading rooms*.

Surry

Surry has a total mean CDF of 1.96E-4 per reactor-year. The calculated annual CDF due to fire is 1.1E-5, which is approximately 6 percent of the total CDF. Fires in the *emergency switchgear room, main control room, auxiliary building*, and *cable vault and tunnel* are important contributors to the fire CDF.

Oconee

The total mean CDF for Oconee is 2.5E-4 per reactor-year. The fire contribution to the mean annual CDF is 1.0E-5 per reactor-year. The fire-induced sequences at Oconee contribute about 4 percent to the total CDF. The fire analysis identified one critical area, the *cable shaft*, which contains virtually all the control cables for the plant systems of importance.

South Texas Project

The total mean CDF is 4.4E-5. The annual CDF due to fire is 4.9E-7, or about 1 percent of the total CDF. As stated, only the *control room* makes a significant contribution to the 1 percent contributed by fire.

Catawba Units 1 and 2

The total mean CDF for Catawba Units 1 and 2 is 7.8E-5 per reactor-year. The calculated annual CDF due to fire is approximately 3.4E-7, which is less than 1 percent of the total. The dominant sequences postulate a fire in either the *control room* or *cable room* that fails the control circuits of redundant trains of equipment.

McGuire

The McGuire IPE estimated a total mean CDF of 7.4E-5 per reactor-year. The calculated annual CDF attributable to fire is approximately 8.1E-8 or less than 1 percent of the total. Major fire sequences involve the *control room* or *cable room* where fires are assumed to fail the control circuits of redundant trains of equipment.

B.3 DETAILED REVIEW OF A BOILING-WATER REACTOR FIRE PRA

The boiling-water reactor (BWR) plant chosen for a detailed review is Peach Bottom, and the resource documents are NUREG-1150 and NUREG/CR-4550 (Volume 4, Part 3).

B.3.1 Internal Events

The total mean CDF from internal events is 4.50E-6 per reactor-year. Station blackout (SBO) contributes to this value with 2.2E-6, that is 48.9 percent of the total CDF. The SBO initiating frequency, from the internal events study, is 0.079, which was taken from WASH-1400 (U.S. Atomic Energy Commission, 1975).

B.3.2 External Events

The overall fire-induced CDF for Peach Bottom Unit 2 is 1.95E-5 per reactor-year. The dominant contributing plant areas are the (1) control room, (2) emergency switchgear room 2C, and (3) emergency switchgear room 2B. These three areas constitute 75 percent of the total fire risk. In the case of the control room, a general transient occurs with smoke-induced abandonment of the area. Failure to control the plant from the remote shutdown panel results in core damage. For the two emergency switchgear rooms, a fire-induced loss of offsite power and failure of one train of the emergency service water (ESW) occurs. Random failure of the other two ESW trains results in SBO and core damage. Tables B.2 and B.3 summarize the results of the fire analysis. Table B.3 shows that the fire in the control room results in a transient and a reactor scram and that the fires in the emergency switchgear rooms contribute to the SBO initiator.

Table B.2 Dominant Peach Bottom Fire Area Contributors to CDF

Fire Area	CDF/RY			
	Mean	5th Percentile	50th Percentile	95th Percentile
Emergency switchgear room 2A	7.4E-7	4.6E-10	1.6E-7	3.0E-6
Emergency switchgear room 2B	3.6E-6	3.5E-9	2.0E-6	1.3E-6
Emergency switchgear room 2C	4.7E-6	4.2E-9	2.2E-6	1.7E-5
Emergency switchgear room 2D	7.4E-7	4.6E-9	1.6E-7	3.0E-6
Emergency switchgear room 3A	7.4E-7	4.6E-10	1.6E-7	3.0E-6
Emergency switchgear room 3B	7.4E-7	4.6E-10	1.6E-7	3.0E-6
Emergency switchgear room 3C	7.4E-7	4.6E-10	1.6E-7	3.0E-6
Emergency switchgear room 3D	8.1E-7	5.3E-10	1.7E-7	3.3E-6
Control room	6.2E-6	4.2E-10	1.4E-6	8.0E-6
Cable spreading room	6.7E-7	9.1E-9	1.7E-7	2.3E-6
Total	2.0E-5	1.1E-6	1.2E-5	6.4E-5

Table B.3 Dominant Accident Sequence Contributors to CDF

Sequence	Fire Area	Mean CDF/RY
T_1BU_1U	Emergency switchgear room 2A	7.4E-7
	Emergency switchgear room 2B	3.6E-6
	Emergency switchgear room 2C	3.6E-6
	Emergency switchgear room 2D	7.4E-7
	Emergency switchgear room 3A	7.4E-7
	Emergency switchgear room 3B	7.4E-7
	Emergency switchgear room 3C	7.4E-7
	Emergency switchgear room 3D	8.1E-7
$T_3U_1U_2X_1U_3$	Control room	6.2E-6
	Cable spreading room	6.7E-7
$T_1BU_1W_1X_2W_2$ $W_3U_4V_2V_3Y$	Emergency switchgear room 2C	8.1E-7
$T_1BU_1W_1X_2W_2$ $W_3U_4V_2Y$	Emergency switchgear room 2C	2.7E-7

Detailed Description of Fire Scenarios in Areas That Are Main Contributors

Control Room

Two scenarios in the control room remained after screening; both are based on a single transient sequence ($T_3U_1U_2X_1U_3$). Both of these scenarios assume abandonment of the control room because of smoke from fire in a cabinet. Credit was given for extinguishing the fire in the burning cabinet quickly, since the control room is continuously staffed. None of the three control room fires in the database led to abandonment of the control room. It was assumed that only 1 in 10 fires would not be extinguished before sufficient smoke was generated to force abandonment of the control room.

This factor (f_R) was taken to be the best estimate of a maximum entropy distribution. As an upper bound, it was assumed that the next control room fire that occurred would force abandonment, and thus, the probability would be 1 in 4. As a lower estimate it was assumed that only 1 in 100 control room fires would lead to abandonment. The Sandia large-scale enclosure tests (NUREG/CR-4527, Vol. 2) have demonstrated that smoke engulfed a mocked-up control room because of a cabinet fire within 6 to 8 minutes from time to ignition, even with ventilation rates of up to 10 room changes per hour. Therefore, these estimates on abandonment probability given a cabinet fire are deemed to be reasonable.

Because of the cabinet configuration within the Peach Bottom control room and considering the Sandia cabinet fire tests, the postulated fire was assumed not to spread or damage any components outside of the cabinet in which the fire started. All control room cabinets at Peach Bottom had penetrations through the cabinet bottom to the cable spreading room below. Also, these cabinets had enclosed backs and tops. In Sandia's cabinet fire tests, cabinets had open backs and enclosed tops. Even in this configuration, fire did not spread to adjacent cabinets. Therefore, the cabinet area ratio factor (f_A) was considered to be known fairly accurately. As a lower bound, it was assumed that only one-half of the applicable cabinet could initiate a sufficiently large fire. An upper bound estimate assumed that all cabinet areas could initiate the fire, but also assumed that a transient fire at a maximum of 1 ft (0.3 m) away from the cabinet in all exposed directions could cause the same damage to the cabinet and the same release of smoke. In both control room scenarios, the fire was assumed to totally disable the functions of the cabinet in which the fire started.

Both fire scenarios assumed that the remote-shutdown system was independent of the control room. This assumption is potentially not - conservative, because the possibility exists that subtle interactions between the remote shutdown panels and the control room are still present. As

part of the Fire Risk Scoping Study (NUREG/CR-5088), an exhaustive cable tracing effort yielded a number of subtle interactions between one plant's control room and the remote-shutdown panel.

Area ratios for fire involvement only considered total cabinet area in the control room. This is based on fire data, which illustrate that the only control room fires to date have occurred in control cabinets.

Control Room Fire Scenario 1: The first scenario postulates a fire starting inside the reactor core isolation cooling (RCIC) cabinet and subsequent smoke release forcing abandonment of the control room. Procedures require that the reactor be manually scrammed, thus a T_3 transient sequence arises. The RCIC system (U_2) is not independent of the control room, since it is not part of the remote shutdown system and is assumed to fail, given a fire in its control cabinet. The control rod drive (CRD) system (U_3) is also not part of the remote shutdown system and, thus, no credit is given for its utilization. The high-pressure coolant injection (HPCI) system (U_1) and the automatic depressurization system (ADS)(X_1) are part of the remote shutdown panel but are failed due to operator error.

The core-damage equation is as follows:

$$\phi_{CM} = \lambda_{CR} \, f_A \, R_{op} \, f_R$$

where:

ϕ_{CM} = fire-induced CDF for control room Scenario 1

λ_{CR} = frequency of control room fires

f_A = area ratio of the RCIC cabinet to total cabinet area within the control room

R_{op} = probability that operators will fail to recover the plant from the remote shutdown panel

f_R = probability that smoke will force abandonment of the control room given a fire

Table B.4 gives the values of each of these factors as well as their associated distribution and upper and lower bounds. For all lognormal and gamma distributed variables in Table B.4 and the following tables, the lower bound and upper bound represent the 5th and 95th percentiles of the distribution, respectively, while the best estimate represents the mean value.

Control Room Fire Scenario 2: The second fire scenario in the control room assumes that the fire is initiated in any cabinet other than the RCIC cabinet. As in the first scenario, subsequent smoke release forces abandonment of the control room. Credit is given for the RCIC system automatically cycling to control reactor level even though it is not controlled from the remote shutdown panel. Therefore, the RCIC system (U_2) must randomly fail, which adds the Q_{RCIC} term in the core damage equation. As in the first scenario, the reactor is manually scrammed (T_3) and the HPCI system (U_1) and ADS (X_1) are failed because of operator error at the remote shutdown panel. Also, no credit is given for the CRD system (U_3), since it is not part of the remote shutdown panel.

The core-damage equation is as follows:

$$\phi_{CM} = \lambda_{CR} (1-f_A) R_{op} Q_{RCIC} f_R$$

where:

ϕ_{CM} = fire-induced CDF for control room Scenario 2

λ_{CR} = frequency of control room fires

$(1-f_A)$ = area ratio of all cabinets other than RCIC cabinet to total cabinet area within the control room

R_{op} = probability that operators will fail to recover the plant from the remote shutdown panel

Q_{RCIC} = random failure of the RCIC system (failure not related to fire)

f_R = probability that smoke will force abandonment of the control room given a fire

Table B.5 gives the values of each of these factors, as well as their associated distribution and upper and lower bounds.

Switchgear Rooms

As mentioned earlier, fires in switchgear rooms 2C and 2B are important contributors and lead to SBO scenarios. The discussion that follows presents a fire scenario in other switchgear rooms. The next sections then present the analysis for switchgear rooms 3D, 2B, and 2C.

Emergency Switchgear Rooms 2A, 2D, 3A, 3B, and 3C: For all five of these fire areas, the scenario is similar. This sequence (T_1BU_1) was an SBO caused by a fire-induced loss of offsite power (T_1) and a random loss of the emergency service water (ESW) system. This random (failure not related to the fire itself) loss of ESW caused an SBO because ESW provides cooling for all four diesel generators. Thus, the emergency onsite power system (B) failed. ESW also provides room cooling for the HPCI system (U_1). The HPCI system will fail in approximately 10–12 hours because of loss of room cooling or because of battery depletion caused by the SBO.

These areas are all similar in that the primary source of fire is electrical switchgear within the fire area. Therefore, the fire frequency was developed for electrical switchgear rooms, and area ratios were for only the cabinet area within the room. A valid mechanism for spread of fire outside these cabinets was required to develop a hot gas layer which would fail offsite power. A plant-specific look at these switchgears showed that in the case of all breaker cubicles, many small cables passed through the top at one penetration and, furthermore, that this penetration was inadequately sealed. There are ventilation slots at the bottom of the cabinets; therefore, given a fire, a chimney effect could occur and it was assumed that there would be a 50-percent chance of the fire exiting the top. Furthermore, a cable run exists directly above these penetrations, which would add more fuel to the fire.

Table B.4 Control Room Fire Scenario 1—Factors and Distributions

Factor	Distribution	Lower Bound	Best Estimate	Upper Bound
λ_{CR}	Gamma	1.2E-7	2.33E-3	6.2E-3
f_A	Maximum entropy	0.01	0.02	0.028
R_{op}	Maximum entropy	6.4E-3	6.4E-2	0.64
f_R	Maximum entropy	0.01	0.1	0.25

Table B.5 Control Room Fire Scenario 2—Factors and Distributions

Factor	Distribution	Lower Bound	Best Estimate	Upper Bound
λ_{CR}	Gamma	1.2E-7	2.33E-3	6.2E-3
$(1-f_A)$	Maximum entropy	0.49	0.98	1.0
R_{op}	Maximum entropy	6.4E-3	6.4E-2	0.64
f_R	Maximum entropy	0.01	0.1	0.25

Since this fire scenario requires that the cable run directly above the 4160-V switchgear ignites to add sufficient fuel to form a hot gas layer within the entire room which then fails offsite power trunks J57 and J58, the area ratio factor (f_A) was the ratio of 4160-V switchgear area to total cabinet area within the fire area. A measurement of this ratio yielded a best estimate of 0.9 for this maximum entropy variable. As a lower bound, only the centermost cubicle was postulated to be capable of failing offsite power and thus, an area ratio of 0.1 was assessed. For an upper bound, it was assumed that the most probable source of fire was the high-voltage 4160-V cubicles and not the other lower voltage cabinet. This led to an upper bound of 1.0. The percentage of cabinet fires (f_S) that would be large enough to exit the top of a cubicle was felt to equal approximately unity on the basis of Sandia fire testing experience. Thus, a tight maximum entropy distribution for the severity ratio factor was postulated. The percentage of fires $Q(\tau_G)$ that are manually extinguished before requisite damage occurs was evaluated previously in that study. The term that represents random failure of the ESW system (Q_{ESW}) can be represented by the following: failures of the emergency diesel generators, a failure to recover one diesel generator within 16 hours, a failure to manually align emergency

service water, and common-cause failures of certain ESW air-operated valves (AOVs).

These failures were developed as part of the internal events analysis of Peach Bottom and are identical except for the postulated mission time of the emergency diesel generators (DGHWNR16HR). A 16-hour mission time was assumed for the diesel generators because offsite power trunks J57 and J58 were irrecoverably lost due to fire damage. Peach Bottom SBO procedures specify, given failure of the emergency diesel generators, that portable generators be transported to the site. It is felt that within 24 hours a portable generator will be in place and cabling will be run to provide some core cooling and, thus, prevent core damage. Failure of the diesel generators at 16 hours and subsequent boiloff from the core would lead to core damage in approximately 24 hours if portable power and core cooling were not in place.

The core-damage equation is as follows:

$$\phi_{CM} = \lambda_{SGR}\, f_A\, f_S\, Q(\tau_G)\, Q_{ESW}\, f_R \qquad (B\text{–}3)$$

where:

ϕ_{CM} = fire-induced CDF for each of the five switchgear rooms 2A, 2D, 3A, 3B, and 3C

λ_{SGR} = frequency of switchgear room fires

f_A = ratio of 4160-V switchgear to total cabinet area within the fire area

f_S = percentage of cabinet fires that would be large enough to exit the top cubicle

$Q(\tau_G)$ = percentage of fires that are not manually extinguished before requisite damage occurs

Q_{ESW} = random failure of the emergency service water system

f_R = percentage of fires that exit the top of a switchgear cubicle

Table B.6 gives the values of each of these factors as well as their associated distribution and upper and lower bounds.

Emergency Switchgear Rooms 3D and 2B: The scenario is identical to the one described previously. However, some fire-related failures of the ESW also occur. For emergency switchgear room 3D, the fire fails power to the ECW pump, while for room 2B, power is failed to ESW pump A. These fire-related failures, coupled with additional random failures, lead to a loss of the ESW system and subsequent SBO.

The only modification to core damage equation B.1 would be to the Q_{ESW} term. For emergency switchgear room 3D, Q_{ESW} consists of failures of two emergency diesel generators, a failure to

recover one train of emergency ac power within 16 hours, and the common-cause failure of selected ESW AOVs; for emergency switchgear room 2B, Q_{ESW} requires an ESW check valve failure in addition to the failures described above for room 3D.

Emergency Switchgear Room 2C: Three scenarios survived screening for this fire area. The first was the SBO scenario described before with fire-related failure of offsite power and ESW pump B. For the other two sequences, SBO does not occur and other random failures lead to long-term core damage scenarios. The core damage equation for all three scenarios is identical to that discussed for emergency switchgear room 2A, except Q_{ESW} is replaced with Q_{RANDOM} for the latter two long-term sequences to reflect that different random failures are necessary to lead to core damage.

Scenario 1: In this case, the Q_{ESW} term is similar to that for ESW room 2B.

Scenario 2: Scenario 2 is a long-term (approximately 30-hour) core damage sequence. The HPCI system (U_1) and low-pressure coolant injection (LPCI) system (V_3) succeed, but core damage eventually occurs because of failure of all modes of the residual heat removal (RHR) system (W_1, W_2, W_3). Fire-related failures are to offsite power, 4160-V ac bus C, and indirectly to 24-V ac bus C. This fire-induced damage fails the suction path logic to the shutdown cooling (SDC) system (W_2) and one of two injection paths for the suppression pool cooling (SPC) system (W_1) and the containment spray (CS) system (W_3). Additional random failures to the emergency diesel generator fail the other injection path for the SPC and CS systems. Containment venting (Y) is failed by loss of the instrument air system cooling and, given a loss of offsite power, the

**Table B.6 Emergency Switchgear Rooms Fire Scenario—
Factors and Distributions**

Factor	Distribution	Lower Bound	Best Estimate	Upper Bound
λ_{SGR}	Gamma	5.8E-7	2.7E-3	5.7E-3
f_A	Maximum entropy	0.1	0.9	1.0
f_S	Maximum entropy	6.4E-3	6.4E-2	0.64
$Q(\tau_G)$	Maximum entropy	0.52	0.77	1.0
f_R	Maximum entropy	0.05	0.5	1.0

turbine building cooling water (TBCW) system is failed. The alternate cooling system, reactor building cooling water (RBCW), is never aligned because of random failure RBC-XHE-FO-SWCH. The CRD system (U_4) is also failed because of a failure to switch cooling.

The terms λ_{SGR}, f_R, f_S, $Q(\tau_G)$, and their associated distributions are identical to the scenario described for emergency switchgear rooms 3D and 2B.

The term Q_{RANDOM} consists of various failures of emergency diesel generator D in conjunction with a switch failure that precludes critical RBCW system alignments.

The core-damage equation is as follows:

$$\phi_{CM} = \lambda_{SGR}\, f_A\, f_S\, Q(\tau_G)\, f_R\, Q_{RANDOM}$$

where all factors are as previously defined.

Table B.6 gives the values of each of the terms as well as their associated distributions.

Scenario 3: As was the case for Scenario 2, long-term (approximately 30-hour) core damage occurs. The HPCI system (U_1) and the low-pressure core spray (LPCS) system (V_2) succeed, but core damage eventually occurs because of failure of all decay heat removal modes of the RHR system (W_1, W_2, W_3). The CRD system (U_4) and containment venting system (Y) fail for reasons identical to those in Scenario 2. However, fire-related damage to emergency bus C fails one injection side of the SPC, CS, and SDC systems, and random failures fail the other injection path. The core damage equation is identical to that for Scenario 2. The only modification is the equation for the term Q_{RANDOM}. In this scenario, Q_{RANDOM} consists of the same RBCW switch failure plus failures of RHR train B.

B.3.3 Conclusion

The Peach Bottom fire risk results present a picture reasonably similar to the internal events and seismic results. The fire-induced CDF is dominated by fire damage to the emergency service water system in conjunction with random failures, coupled with fire-induced loss of offsite power. In all eight emergency switchgear rooms (four for both Units 2 and 3), both trains (J57 and J58) of offsite power are routed. In each of these areas, breaker cubicles for the 4.1-kV switchgear have a penetration at the top, which has many small cables routed through it. These penetrations are inadequately sealed, allowing the fire to spread to cabling that is directly above the switchgear. This cabling is a sufficient fuel source for the fire to cause a rapid formation of a hot gas layer, which would then lead to a loss of offsite power. Since both offsite power and the emergency service water systems are lost, a station blackout would occur, which would also fail all containment heat removal. A number of possible modifications can be envisioned, including the following:

- more adequate seals for the penetrations on top of the 4.1-kV switchgear cubicles

- spraying fire retardant on the cabling located directly above 4.1-kV switchgear

B.4 DETAILED REVIEW OF A PRESSURIZED-WATER-REACTOR PRA

The pressurized-water-reactor (PWR) plant chosen for a detailed review is Surry Unit 1, and the resource documents are NUREG-1150 and NUREG/CR-4550, Volume 3, Part 3.

B.4.1 Internal Events

The total mean CDF from internal events is 4.0E-5. Station blackout (SBO) contributes to this value with 2.74E-5, that is, 68.5 percent of the total CDF. The SBO initiating frequency, from the internal events study, is 7.0E-2, which was taken from NUREG-1032.

B.4.2 External Events

The overall fire-induced CDF for Surry Unit 1 is 1.13E-5 per reactor-year. The dominant contributing plant areas are the (1) emergency switchgear room, (2) auxiliary building, (3) control room, and (4) cable vault/tunnel. These four areas comprise 99 percent of the total fire risk. In the case of the emergency switchgear room, cable vault/tunnel, and auxiliary building, a reactor coolant pump seal loss-of-coolant accident (LOCA) leads to core damage. The fire itself fails cabling for both the high-pressure injection (HPI) and component cooling water (CCW) systems, resulting in a seal LOCA. For the control room, a general transient with a subsequent stuck-open power-operated relief valve (PORV) leads to a small LOCA. Failure to control the plant from the auxiliary shutdown panel results in core damage. Tables B.7 and B.8 summarize the results of the fire analysis. Table B.8 shows that the main contributors are

- emergency switchgear room (6.09E-6)
- auxiliary building (2.18E-6)
- control room (1.58E-6)
- cable vault/tunnel (1.49E-6)

Table B.8 shows that fires in all four of the main areas contribute to the transient initiator.

Detailed Description of Fire Scenarios in Areas That Are Main Contributors

Auxiliary Building

One fire scenario in the auxiliary building remained after screening. This scenario was a large fire on the 13-ft elevation that irrecoverably damaged power or control cables for both the HPI

Table B.7 Dominant Surry Fire Area Contributors to CDF

Fire Area	CDF/RY			
	Mean	5th Percentile	Median	95th Percentile
Emergency switchgear room	6.09E-6	3.93E-9	3.15E-6	1.98E-5
Control room	1.58E-6	1.20E-10	4.68E-7	6.95E-6
Cable vault/tunnel	1.49E-6	6.51E-10	6.99E-7	5.79E-6
Auxiliary building	2.18E-6	5.32E-7	1.59E-6	5.64E-6
Charging pump service water pump room	3.92E-8	1.43E-10	5.66E-9	1.58E-7
Total	**1.13E-5**	**5.37E-7**	**8.32E-6**	**3.83E-5**

Table B.8 Dominant Accident Sequence Contributors to CDF

Sequence	Fire Area	Mean CDF/RY
$T_3D_3WD_1$	Emergency switchgear room Auxiliary building Cable vault/tunnel	6.09E-6 2.18E-6 1.49E-6
T_3QD_1	Control room Charging pump service water pump room	1.58E-6 3.92E-8

and CCW systems. These fire-related failures with no additional random failures required led to a reactor coolant pump seal LOCA. The recovery for this particular scenario required the operation of two manual HPI system cross-connect valves located in the immediate vicinity of the large fire. No recovery was allowed until 15 minutes after the fire was extinguished.

The core-damage equation is as follows:

$$\phi_{CM} = \lambda_{aux}\, f_A\, f_S\, Q(\tau_G)\, R_{op}$$

where:

ϕ_{CM} = fire-induced CDF for the auxiliary building

λ_{aux} = frequency of auxiliary building fires

f_A = area ratio within the auxiliary building where critical damage occurred

f_S = severity ratio (based on generic combustible fuel loading) for a large fire

$Q(\tau_G)$ = percentage of fires in the suppression database that were not manually extinguished before the COMPBRN-predicted time to critical damage occurred

R_{op} = failure to cross-connect the Unit 2 HPI system to either prevent a seal LOCA or mitigate its effect

Table B.9 gives the values of each of these factors as well as their associated distribution and upper and lower bounds. For all lognormal distributed variables in Table B.9, the lower bound and upper bound represent the 5th and 95th percentiles of the distribution, respectively, while the best estimate represents the mean value.

Cable Vault/Tunnel

The one remaining scenario that survived screening is similar to the one described for the auxiliary building in that the postulated fire irrecoverably damages power or control cables for both the HPI and CCW systems, leading to a seal LOCA.

Credit was taken for the automatic carbon dioxide (CO_2) system suppressing the fire before critical damage occurred. COMPBRN predicted 3 minutes' time to damage for this particular scenario. The automatic CO_2 system is actuated by fixed-temperature heat detectors at 190 °F (361 K). One heat detector is located at the end of the critical area of influence for this scenario. Two others are located so that ventilation flow would force the hot gas layer in their direction. The system actuation delay time to allow for evacuation is 30 seconds. Therefore, the heat detectors must respond to fire ignition and the CO_2 system must suppress the fire within 2.5 minutes to prevent critical damage. For these reasons, system reliability data for automatic CO_2 suppression systems were modified to account for this relatively short time to prevent critical damage.

Operator recovery for this scenario is similar to that for the auxiliary building scenario, except that the fire is not in the immediate vicinity or even in the same fire area in which the local recovery actions must take place. Also, since no control

Table B.9 Auxiliary Building Fire Scenario—Factors and Distributions

Factor	Distribution	Lower Bound	Best Estimate	Upper Bound
λ_{aux}	Gamma	0.027	0.066	0.16
f_A	Maximum entropy	2.4E-4	6.3E-4	1.1E-3
f_S	Maximum entropy	0.19	0.30	0.67
$Q(\tau_G)$	Maximum entropy	0.69	0.80	1.0
R_{op}	Maximum entropy	0.19	0.26	1.0

room operators respond to the fire itself, the same recovery value for operator action was applied as was used in the internal events analysis.

The core-damage equation is as follows:

$$\phi_{CM} = \lambda_{CSR} \, f_A \, f_S \, Q(\tau_G) \, Q_{AUTO} \, R_{op}$$

where:

ϕ_{CM} = fire-induced CDF for the cable vault/tunnel

λ_{CSR} = frequency of cable vault/tunnel fires

f_A = area ratio within the cable vault/tunnel where critical damage occurred

f_S = severity ratio (based on generic combustible fuel loading)

$Q(\tau_G)$ = percentage of fires in the database that were not manually extinguished before the COMPBRN-predicted time to critical damage occurred

Q_{AUTO} = probability of the automatic CO_2 system not suppressing the fire before the COMPBRN- predicted time to critical damage occurred

R_{op} = failure to cross-connect the Unit 2 HPI system to either prevent a seal LOCA or mitigate its effect

Table B.10 gives the values of each of these factors, as well as their associated distribution and upper and lower bounds.

Control Room

One scenario survived the screening process for the control room. As was the case for the auxiliary building and cable vault/tunnel, no additional random failures were required to lead directly to core damage. This scenario was a fire interior to benchboard 1-1 leading to the spurious actuation of one PORV located on this benchboard. Because of the cabinet configuration within the control room and considering Sandia cabinet fire tests, the fire was assumed not to spread or damage any components outside of benchboard 1-1. However, because of the Sandia large-scale enclosure tests (where smoke engulfed a control room within 5–10 minutes of time from ignition within a cabinet even with ventilation rates of up to 10 room changes per hour), this scenario postulates forced abandonment of the control room and subsequent plant control from the auxiliary shutdown panel located in the emergency switchgear room.

Credit was given for extinguishing the fire quickly within benchboard 1-1, since the control room is continuously staffed. None of the four control room fires in the database led to abandonment of the control room. It was assumed that 10 percent of all control room fires would result in abandonment of the control room, and a factor of 10 reduction in control room fire frequency was the modification made to allow credit for continuous occupation.

Table B.10 Cable Vault/Tunnel Fire Scenario—Factors and Distributions

Factor	Distribution	Lower Bound	Best Estimate	Upper Bound
λ_{CSR}	Gamma	3.0E-6	7.5E-3	0.016
f_A	Maximum entropy	0.011	0.025	0.047
f_S	Maximum entropy	0.50	0.99	1.0
$Q(\tau_G)$	Maximum entropy	0.69	0.80	1.0
Q_{AUTO}	Maximum entropy	0.50	0.70	0.90
R_{op}	Maximum entropy	4.4E-3	0.044	0.44

The area ratio for fire involvement was developed by comparing the area of benchboard 1-1 to the total cabinet area in the control room. This is warranted because fire event data show that all control room fires have occurred within electrical cabinets. Therefore, this is postulated to be the most likely fire ignition source within the control room.

Once the control room is abandoned, operators would control the plant from the auxiliary shutdown panel. However, PORV indication is not provided at this panel and in conversations with the utility it was learned that the PORV "disable" function on the auxiliary shutdown panel is not electrically independent of the control room. Therefore, it was assumed that the PORV disable function would fail and, consequently, the operators would be in a high stress recovery mode.

The core-damage equation is as follows:

$$\phi_{CM} = \lambda_{CR}\, f_A\, R_{op}\, f_R$$

where:

ϕ_{CM} = fire-induced CDF for the control room

λ_{CR} = frequency of control room fires

f_A = ratio of benchboard 1-1 area to total cabinet area within the control room

R_{op} = probability that operator will not successfully recover the plant from the auxiliary shutdown panel

f_R = probability that operators will not successfully extinguish the fire before smoke forces abandonment of the control room

Table B.11 gives the values of each of these factors as well as their associated distribution and upper and lower bounds.

Emergency Switchgear Room

One fire scenario remained for the emergency switchgear room after screening. This scenario was a fire that damaged either power or control cables for HPI and CCW pumps, thus leading to a reactor coolant pump seal LOCA. No additional random failures were required for this scenario to lead directly to core damage.

As was the case for the cable vault/tunnel and auxiliary building, recovery from this scenario was by cross-connecting HPI from Unit 2. The fire itself would not affect local auxiliary building recovery actions. Therefore, similar to the cable vault/tunnel, the same probability for recovery was used as in the internal events analysis.

The core-damage equation is as follows:

$$\phi_{CM} = \lambda_{SGR}\, Q(\tau_G)\, R_{op}\, (f_{A1}\, f_{S1} + f_{A2}\, f_{S2})$$

where:

ϕ_{CM} = fire-induced CDF for the emergency switchgear room

Table B.11 Control Room Fire Scenario—Factors and Distributions

Factor	Distribution	Lower Bound	Best Estimate	Upper Bound
λ_{CR}	Gamma	1.2E-6	1.8E-3	7.4E-3
f_A	Maximum entropy	0.028	0.084	0.12
R_{op}	Maximum entropy	7.4E-3	0.074	0.74
f_S	Maximum entropy	0.01	0.1	0.25

λ_{SGR} = frequency of emergency switchgear room fires

$Q(\tau_G)$ = percentage of fires in the database that were not manually extinguished before the COMPBRN-predicted time to critical damage occurred

R_{op} = failure to cross-connect the Unit 2 HPI system to either prevent a seal LOCA or mitigate its effect

f_{A1} = area ratio within the emergency switchgear room for a small fire where critical damage occurred

f_{S1} = severity ratio (based on generic combustible fuel loading) of small fires

f_{A2} = area ratio within the emergency switchgear room for a large fire

f_{S2} = severity ratio (based on generic combustible fuel loading) of large fires

Table B.12 gives the values of each of these factors, as well as their associated distribution and upper and lower bounds.

B.4.3 Conclusion

The overall fire-induced CDF for Surry Unit 1 is 1.13E-5 per reactor-year. The dominant contributing plant areas are the following: (1) emergency switchgear room, (2) auxiliary building, (3) control room, and (4) cable vault/tunnel. These four areas constitute 99 percent of the total fire risk.

In the case of the emergency switchgear room, cable vault/tunnel, and auxiliary building, a reactor coolant pump seal LOCA leads to core damage. The fire itself fails cabling for both the HPI and CCW systems, resulting in a seal LOCA.

For the control room, a general transient with a subsequent stuck-open PORV leads to a small LOCA. Failure to control the plant from the auxiliary shutdown panel results in core damage.

B.5 REFERENCES

Commonwealth Edison Co., "Zion Station Unit 1 and 2 Probabilistic Safety Study," NRC Docket Nos. 50-295 and 50-304, Zion, Illinois, September 1981.

Consolidated Edison Company of New York, Inc./Halliburton NUS Environmental Corp., "Individual Plant Examination for Indian Point Unit 2 Nuclear Generating Station," Buchanan, New York, August 1992.

Consumers Power Company, "Big Rock Point Plant Probabilistic Risk Assessment," Jackson, Michigan, March 1981.

Table B.12 Emergency Switchgear Room Fire Scenario—Factors and Distributions

Factor	Distribution	Lower Bound	Best Estimate	Upper Bound
λ_{SGR}	Gamma	2.0E-5	8.0E-3	0.017
f_{A1}	Maximum entropy	0.02	0.039	0.099
f_{S1}	Maximum entropy	0.33	0.7	0.81
f_{A2}	Maximum entropy	0.051	0.10	0.24
f_{S2}	Maximum entropy	0.19	0.30	0.67
$Q(\tau_G)$	Maximum entropy	0.67	0.80	1.0
R_{op}	Maximum entropy	4.4E-3	0.044	0.44

Cross, R.B., et al., "South Texas Project Electric Generating Station Level 2 PRA and Individual Plant Examination," Hanston Lighting & Power Company and Pickard, Lowe and Garrick, Inc., Palacios, Texas, August 1992.

Duke Power Company, "Catawba Nuclear Station IPE Submittal Report," Clover, South Carolina, September 1992.

———, "IPE Submittal Report for McGuire Nuclear Station," Cornelius, North Carolina, November 1991.

Garrick, B.J., et al., "Seabrook Station Probabilistic Risk Assessment," Pickard, Lowe, and Garrick, Inc., Framingham, Massachusetts, December 1983.

Nuclear Safety Analysis Center, NSAC60, Vol. 1, "A Probabilistic Risk Assessment of Oconee Unit 3," Palo Alto, California, June 1984.

NUS Corporation, "Severe Accident Risk Assessment, Limerick Generating Station," Pottstown, Pennsylvania, April 1983.

U.S. Atomic Energy Commission, WASH-1400 (now NUREG-75/014), "Reactor Safety Study—An Assessment of Accident Risks in U.S. Commercial Nuclear Power Plants," October 1975.

U.S. Nuclear Regulatory Commission, NUREG-1032, "Evaluation of Station Blackout Accidents at Nuclear Power Plants," P. Baranowski, June 1988.

———, NUREG-1150, Vol. 1, "Severe Accident Risks: An Assessment for Five U.S. Nuclear Power Plants," December 1990.

———, NUREG/CR-4527, Vol. 2, "An Experimental Investigation of Internally Ignited Fires in Nuclear Power Plant Cabinets, Part III—Room Effects Tests," Sandia National Laboratories, Albuquerque, New Mexico, October 1988.

———, NUREG/CR-4550, Vol. 3, Part 3, "Analysis of Core Damage Frequency, Surry Power Station, Unit 1, External Events," Sandia National Laboratories, Albuquerque, New Mexico, December 1990.

———, NUREG/CR-4550, Vol. 4, Part 3, "Analysis of Core Damage Frequency: Peach Bottom Unit 2 External Events," Sandia National Laboratories, Albuquerque, New Mexico, December 1990.

———, NUREG/CR-4832, Vol. 1, "Analysis of the LaSalle Unit 2 Nuclear Power Plant: Risk Methods Integration Program and Evaluation Program (RMIEP), Summary," July 1992.

———, NUREG/CR-5088, "Fire Risk Scoping Study: Investigation of Nuclear Power Plant Fire Risk, Including Previously Unaddressed Issues," Sandia National Laboratories, Albuquerque, New Mexico, January 1989.

Appendix C

FIRE MODELING UNCERTAINTY

CONTENTS

APPENDIX C
FIRE MODELING UNCERTAINTY

As discussed in Chapter 4 and further illustrated in Chapter 6, risk-informed and performance-based approaches may result in the adoption of alternative methods for compliance on a plant-specific basis. The justifications for the use of such alternative methods may come from the following four types of analyses:

(1) Cases in which the equipment contained within a fire compartment may have an insignificant contribution to core damage frequency or risk even if all the equipment is damaged as a result of a single-exposure fire.

(2) Cases in which data analyses and reliability modeling are used to show that the performance of an alternative design is equivalent to or better than the base case. Examples are the relaxation of the surveillance interval and modification of surveillance strategy as discussed in Chapter 6.

(3) Cases in which deterministic analyses (fire modeling) reject or accept a given hypothesis. As an example of the case study on the separation distance between redundant cable trays, it was shown that the redundant cables will not be damaged if they are separated by more than 4.6 m (15 ft) and as long as the peak heat release rate of the fire source is below 2 MW. For this case study, when the performance measure is defined as damage to redundant cable, 4.6-m (15-ft) separation and 6.1-m (20-ft) separation will provide equivalent performance.

(4) Cases in which none of the above three analyses by themselves could result in a justifiable decision; however, if integrated systematically they could provide the necessary justification. In the integrated analyses, the measure estimated is typically the change in core damage frequency (ΔCDF) or risk, and the decision may be

made on the basis of value-impact evaluation of the change in risk and other factors.

Regardless of the analysis type, the following issues regarding the sources of uncertainties need to be addressed:

(1) availability and quality of information about uncertainties in the input variables to the model and in the parameters used in the model

(2) accuracy of the model, excluding any input variability discussed above

The uncertainty distribution, associated with input variables and model parameters (Issue 1), is estimated using measurements or monitored data through application of the Bayes method (Kaplan, 1983). Computer software is widely used for these types of uncertainty analyses for both risk-informed and performance-based models. This technology has been utilized for more than a decade in various probabilistic risk assessments and reliability studies.

The uncertainties in input variables and the model parameters are propagated through an integrated model using Monte Carlo sampling techniques. Variance reduction techniques and stratified sampling strategies have been extensively used to propagate the uncertainties in an efficient manner. Software such as in the IRRAS computer code (NUREG/CR-5813)) and the COMPBRN (EPRI NP-7282)) code have already implemented these techniques for uncertainty propagation. Other methods, such as discrete probability propagation and moment propagation, have been used less extensively.

The accuracy of model prediction (Issue 2), excluding the variabilities of the input and model parameters, is entrenched in code validation. In most cases, simplifying assumptions have been incorporated to reduce the code's development effort and to facilitate the large number of runs usually required for risk-informed and performance-based evaluations.

Two methods of validation are usually proposed. The first is the comparison of the code predictions to those of another validated code that is more comprehensive and suffers from fewer simplifying assumptions. The other method requires comparison of the code predictions to available measurements obtained through a well-instrumented experiment.

In any case, exhaustive comparisons of the existing codes to either experiments or a more comprehensive code are not generally feasible because of the large number of case runs that may be necessary or the cost associated with new experiments and/or additional computer runs. Various statistical methods are available to provide an estimate of the inaccuracies of the code prediction using a small set of validation runs.

Currently, expert judgments are used in most cases to determine the accuracy of the code predictions in light of the limited experimental data available. One method used in the building industry, albeit informal, aggregates the results of those fire experiments (or actual fire events) that are judged to be representative of the case under study, in order to refine the code estimates. The aggregation process is based on the weighted mixture of all results. The closer the fire experiment represents the case run, the higher would be its weight. This is also the case for the computer codes for evaluating fire propagation times.

C.1 INSIGHTS REGARDING THE UNCERTAINTIES IN COMPBRN IIIe

This section contains a preliminary discussion of the potential un-certainties in the COMPBRN code. As discussed in Chapter 4, fire modeling codes have been used in estimating the time it takes for fire to damage critical components if the fire is not suppressed. Also discussed was the fact that a fire modeling code generally simulates two major phenomena. One phenomenon deals with

the strength of the fire source in terms of heat release rate as a function of time, and the other deals with the thermal environment as a result of the fire, including radiation, to the target object to estimate the damage time. Also discussed was that the COMPBRN series is perhaps the only available computer code that attempts to model both phenomena. Other computer codes, such as CFAST and FPETOOL, currently model the thermal transport phenomenon and accept the fire heat release rate as a function of time as input. FIVE (fire-induced vulnerability evaluation) methodology and worksheets are similar to the latter group of codes, but do not model the fire source strength, although some guidelines are provided for simple cases.

The sources of initiating fires in nuclear power plants vary: cable fire, oil fire, transient fire, cabinet fire, etc. Experience accumulated from earlier fires and fire tests show large variability in fire heat release rate even for the same type of fire source. For example, a cabinet fire involving high-voltage equipment is fundamentally different from fires initiated in cabinets containing low-voltage equipment. Earlier fires in nuclear power plants have shown that cabinet fires involving high-voltage equipment generate tremendous amounts of heat, some due to electrical energy converted to thermal as a result of electrical faults. In contrast, a slow, smoldering fire may occur in cabinets containing low-voltage equipment. Also, various tests performed by the Electric Power Research Institute and Sandia National Laboratories have shown that heat-release rate from cable tray fires is a complex phenomenon, depending on many parameters, such as cable orientation, cable location, ventilation, and size of the initiating transient fire.

The heat-release rate of a fire source is a complex physical phenomenon and, given the current state-of-the-art modeling techniques, one may expect large uncertainties associated with the code prediction. Typically, simplified bounding estimation using a surface-controlled burning rate model has been utilized in the computer codes, such as COMPBRN, to ensure the conservative estimation of the fire impact. Because of the conservative nature of such modeling, there may be cases in which the heat release rate is significantly overpredicted. In these cases, the peak heat release rates and the associated ranges of variation (uncertainty) may be subjectively determined in light of past occurrences of fire or fire tests and used as input to the code.

To gain some insights on uncertainty issues regarding fire modeling, the following three cases are discussed:

(1) source fire heat-release rate of a case-specific Institute of Electrical and Electronics Engineers (IEEE) 383 rated cable tray

(2) transport of thermal environment for a given fire source heat-release rate

(3) integrated verification using fire test data

C.1.1 Case 1: Source Fire Heat-Release Rate of a Case-Specific IEEE-Rated Cable Tray

The specific case selected for this analyses is Case 1 described in Section 6.2.1.3 (see Table 6.4), in which the size of the pilot fire is 1.2 m (4 ft) × 0.6 m (2 ft) in the lowest cable tray. Details of this case study is provided in Appendix D. COMPBRN predicted a total heat-release rate, which is given in Table C.1. In an earlier study (NUREG/CR-4230), similar fire scenarios were analyzed using the old version of the COMPBRN code and the conclusion reached was that the heat-release rate predicted by the code was unrealistically high. On the basis of the amount of oxygen available in the plume for the maximum height of the flame, the study concluded that the peak heat-release rate will be limited to 2.5 MW, or about 0.83 MW for a cable tray. On the basis of the fire tests reported in

EPRI NP-2660 and EPRI NP-2751, one may also arrive, for the peak heat-release rate for a cable tray 15 m (~50 ft) long, at a range from 0.9 MW to 2 MW for a well-ventilated room.

To further analyze the availability of oxygen to support the burning rates predicted by COMPBRN, an input deck for the CFAST code was developed. The CFAST computer code is capable of evaluating the concentration of various species of air and combustible products in the hot layer region. According to the CFAST run, at about 5 minutes, the upper hot layer descends to the level of the lowest burning tray. The concentration of oxygen in the hot layer at 5 minutes was estimated to be below 10 percent (ordinary air is 21 percent). Therefore, the heat-release rate will not increase any further because of oxygen depletion and the fire may die down shortly. Accordingly, the peak heat-release rate for this specific case will be below 2 MW and the heat-release rate predicted by COMPBRN after 5 minutes may be overly conservative.

The preceding discussion gives an example of the level of conservatism embedded in the COMPBRN code and shows the role of the analyst in determining the heat-release rate from various sources, considering the complexity and the uncertainties associated with this issue. The heat-release rate is the driving force for the plume mass flow rate, the ceiling jet temperature, and finally, the hot layer temperature which is driven by energy balance. The fire heat-release rate is dependent on the initial fire size, the growth of fire by propagation and ignition of additional combustibles, and the heat-release rate from these additional combustibles. There is a large variability in initial fire size which typically is categorized into three categories—small, medium, and large. The size of fire, associated with each category itself is an uncertain quantity and

Table C.1 Heat-Release Rate (HRR) Predicted by COMPBRN for Case 1 of Safe-Shutdown Distance Case Study

Time (Minutes)	0	1	2	3	4	5	6	7	8	9	10
HRR (MW)	0	0.1864	0.242	0.357	0.895	1.512	2.598	3.472	4.741	7.30	14.06

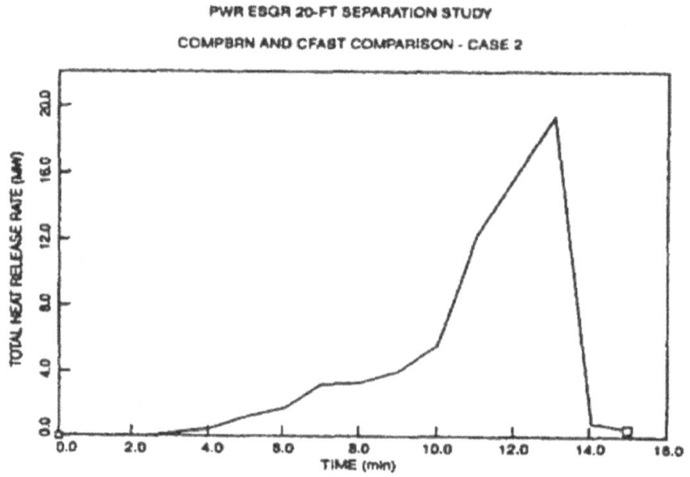

PWR ESGR 20-FT SEPARATION STUDY

COMPBRN AND CFAST COMPARISON - CASE 2

Figure C.1
COMPBRN-Predicted Heat Release From Burning Cables

typically is assigned to some extent subjectively. This initial fire may engulf additional combustibles which result in additional fire growth. The heat-release rate currently is simply calculated in COMPBRN and some of the codes reviewed on the basis of complete burning of vaporized combustible which is empirically measured (surface burning rate). The availability of oxygen, and its impact on limiting the burning rate, sometimes referred to as ventilation-controlled burning or method-of-oxygen-depletion calorimetry, is not typically modeled. This results in a very conservative estimate for burning rates in stacked cable trays that are located near the ceiling.

C.1.2 Case 2: Transport of Thermal Environment for a Given Fire Source Heat-Release Rate

The "transport of thermal environment" routines in the COMPBRN computer code were compared to the CFAST code for Case 2 of the "safe separation distance" case study of Section 6.2.1.3. The oxygen-starvation routine of CFAST was switched off to allow this comparison.

Case 2 is the case in which the size of the pilot fire is reduced to 0.6 m × 0.6 m (2 ft × 2 ft).

Because the pilot fire is not simulated in the CFAST code, the total release rate due to fire predicted by COMPBRN is provided as input to the CFAST code. Using this heat-release rate, CFAST predicted temperatures of the hot gas layer and the target cable tray, and the height of the hot gas layer are compared with those predicted by COMPBRN.

Figure C.1 illustrates the COMPBRN-estimated heat-release rate. According to the COMPBRN code, the heat released by a burning fuel element is determined by three parameters: combustion efficiency, heat of combustion, and mass burning rate. The first two parameters are user-specified input data. (The values used in the present analysis are 0.7 and 0.265E8 J/kg (~11,400 Btu/lb) for the two parameters, respectively.) Since the forced ventilation model is not used in the present study, the mass burning rate is governed by the fire surface area, a specific burning rate constant (0.43E-2 kg/m^2-sec (8.9E-4 lb/s^2/ft)), and a surface-controlled burning rate constant (0.4E-6 kg/J (0.001 lb/Btu)). Because of the simplified physical model and the requirement of several user-controlled input parameters, the COMPBRN-estimated heat-release rate may involve some degree of uncertainty as just discussed.

Comparisons of hot gas layer temperatures are shown in Figure C.2, which shows that the two predictions agree very well for the first 6 minutes. After the COMPBRN-predicted ignition of tray C2 at 5 minutes and tray B (the target tray) at 10 minutes, the hot gas layer temperature predicted by COMPBRN is much higher than that predicted by CFAST. CFAST predicted a much lower elevation of the hot/cold gas layer interface than did COMPBRN. The thicker hot gas layer probably contributes to the lower temperature in the hot region predicted by CFAST.

Figure C.3 compares the target cable tray temperatures. In the COMPBRN analysis, the target is located at an elevation of 4.27 m (9.4 ft) above the floor. Because it is within the hot gas layer region, it receives radiative and convective heat transfer and is heated up continuously as illustrated in Figure C.3. The target reaches the user-specified ignition temperature of 733 K (860 °F) at about 10 minutes, from which time the target remains at the constant ignition temperature. In the CFAST analysis, the target is located at an elevation of 3 m (9.8 ft) from the floor, the same elevation as the pilot fire source. CFAST shows that the target temperature increases from an initial 300 K (81 °F) to about 437 K (327 °F) during the first 5 minutes when the target is outside the hot layer and the dominant heat transfer mechanism is radiation heating. Because CFAST does not model the ignition of the target cable tray, the target temperature increases continuously and reaches a peak of about 957 K (1263 °F) at 13 minutes. In general, the target temperature follows the heat-release rate given in Figure C.1.

Finally, the CFAST code has an option to terminate fire growth if sufficient oxygen is not available in the room. This option was not used in the present comparison study.

Uncertainties of COMPBRN IIIe

The quasi-static two-zone approach used in fire models such as in COMPBRN code involves a large degree of uncertainty in simulating the process of fire growth. To address the uncertainties, the code provides many user-

specified input parameters that can be adjusted to perform uncertainty or sensitivity studies. These parameters include physical property data for combustible materials, model parameters, and variability factors.

The physical property data are needed to define the behavior of the fuel. Table C.2 gives the 14 property parameters required as input to the code. Some of the properties, such as heat value, damage and ignition temperatures, specific burning rate constant, reflectivity, and absorption coefficient, are significant in the damage time assessment. A reasonable estimate of these parameters is essential for the COMPBRN analysis.

The model parameters used as inputs are needed to represent the uncertainties of the simplified physical models in the code. These parameters are related to the physical modeling of heat transfer, forced ventilation, doorway, enclosure walls, and flame/plume entrainment. The seven model parameters and the COMPBRN-suggested values are given in Table C.3. Many of the suggested values were determined by comparisons with experiments.

The variability factors are provided by the code to allow users to multiply the results of various models by a specified modification factor. Since these factors are introduced to modify such values as the burning rate, flame height, heat transfer, and temperature, they are expected to be able to play an important role in the assessment of fire growth. The 14 variability factors and their default values are listed in Table C.4. All default values were used in the present analysis.

C.1.3 Case 3: Integrated Verification of COMPBRN III and COMPBRN IIIe Using Fire Test Data

Verification of the COMPBRN III code is described by Ho et al. (1988) and is important because the code utilizes approximations that go beyond some of the other two-layer codes. These include not calculating the heat loss to the walls of the compartment, but instead assigning a fraction of the heat of combustion to the loss

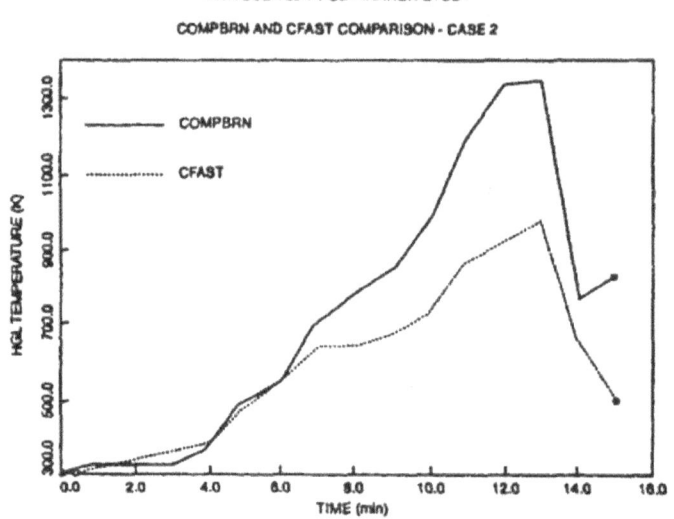

Figure C.2
Comparison of Hot Gas Layer Temperatures

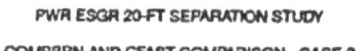

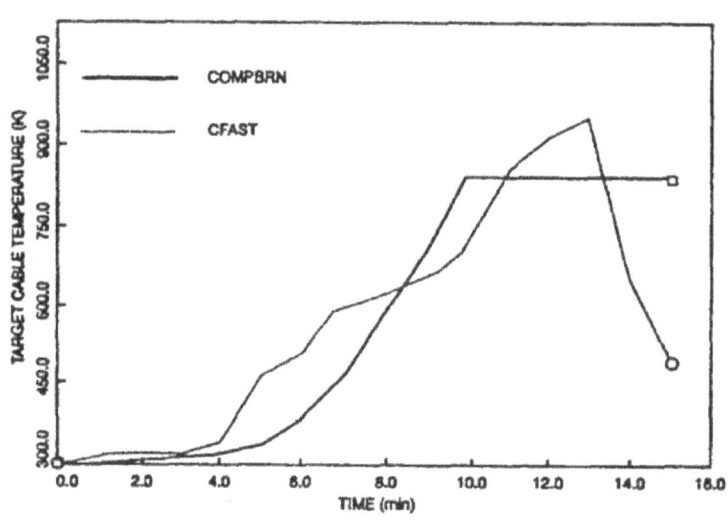

Figure C.3
Comparisons of Target Cable Tray Temperatures

Table C.2 Physical Property Parameters

Property Parameters	Suggested Values	
	Cable	Oil
Density, kg/m³	1710	900
Specific heat, J/kg/K	1040	2100
Thermal conductivity, W/m/K	0.092	0.145
Heat value, MJ/kg	20.6	46.7
Pilot ignition temperature, K	773	400
Spontaneous ignition temperature, K	776	486
Damage temperature, K	623	-
Ventilation-controlled burning rate constant	0.11	0.11
Specific burning rate constant, kg/m²-sec	0.0043	0.061
Surface control burning rate constant, kg/J	0.18×10^{-6}	0.2×10^{-6}
Combustion efficiency	0.7	0.9
Fraction of flame heat released as radiation	0.4	0.45
Absorption coefficient for flame gases, 1/m	1.4	1.4
Reflectivity	0.2	0.35

Table C.3 Model Parameters

Model Parameters	Suggested Value
Heat transfer coefficient for heat transfer in a flame, W/m²/K	22
Convective heat transfer coefficient outside of hot gas layer, W/m²/K	10
Coefficient of inflow air through doorway	0.6
Coefficient of discharge for doorway	0.7
Absorption coefficient of hot gas (1/m)	1.3
Heat transfer coefficient for ceiling and for objects in the hot gas layer, W/m²/K	10
Buoyant plume entrainment coefficient*	2.0

*The buoyant plume entrainment coefficient
= 2.0 for pool fire unaffected by enclosure
= 1.5 for pool fire next to a wall
= 1.25 for pool fire at a corner

Table C.4 User-Specified Variability Factors in COMPBRN IIIe

Variability Factors	Default Value
Ventilation-controlled burning rate	1.0
Fuel-surface-controlled burning rate	1.0
Flame height for horizontal fuel	1.0
Flame height for vertical fuel	1.0
Radiative heat flux interchange	1.0
Buoyant plume temperature	1.0
Convective heat transfer coefficient for vertical surface in plume	1.0
Convective heat transfer coefficient for horizontal surface in plume	1.0
Gas layer local temperature	1.0
Heat transfer to self for vertical fuel	1.0
Heat transfer to adjacent fuel	1.0
Heat flux from ceiling hot gas layer	1.0
Heat flux from re-radiation from walls and barriers	1.0
Mass burnout fraction	1.0

(mostly by radiation) from the plume, and assuming that the fire burns typically through surface-controlled burning with a specified combustion efficiency. Gas concentrations are not calculated. These simplifying assumptions have an important benefit, as described in EPRI NP-7282. The program runs very fast, making it feasible to assign distribution functions to the imprecisely known input variables, and to make multiple runs to obtain a Monte Carlo distribution of the results for use in risk analyses. It is important to evaluate how well the program is able to predict the environmental parameters important in a nuclear power plant compartment fire. The verification process (Ho et al., 1988) used two sets of data from the literature. The first set (Steckler et al., 1984) involved carefully instrumented tests using a constant methane burner fire to cause buoyancy-driven flows out of and into a doorway. The results were characterized by calculated inflow and outflow coefficients.

When the same doorway was used with different rates of burner heat release, the inflow coefficients varied from 0.73 to 1.60 and the outflow coefficients varied from 0.69 to 0.90. COMPBRN III closely reproduced the experimental upper layer temperature, layer height in the room, and layer height in the doorway, if the correct doorway coefficients were used (Ho et al., 1988). This is more than a simple demonstration that COMPBRN III does the arithmetic correctly, however, because a two-layer model had not been used for the experiment (Steckler et al., 1984). Rather, the flows into and out of the doorway were integrated according to the measured temperature profiles. Ho et al. (1988) assume for these runs that the fraction of the heat of combustion lost by radiation is 0.15, which is reasonable for a methane flame that produces no soot.

Data from NUREG/CR-3192 are used to test the capability of COMPBRN III to predict hot gas layer temperature, heat flux to cables in a cable tray at a 6.1-m (20-ft) distance from a pan of burning heptane, and temperatures of the cables. The burning rate, in some tests, did change with time because, depending on the size of the doorway, the air in the room could become vitiated with oxygen. Figures C.4 through C.7 show how the layer temperature, heat flux, and cable jacket temperature vary with the four different doorway sizes and show the COMPBRN III predictions for various assumed combustion efficiencies. In all cases, the assumed fraction of heat lost from the plume by radiation was 0.4, a reasonable number for a flame-producing soot. A combustion efficiency of 0.85, also a reasonable value for a liquid pool fire, seems a good average. The results of verification indicate good agreement between the test and the code prediction.

Note that for both sets of verification data, the situation is consistent with the assumptions of COMPBRN III. The fires were reasonably constant at heat-release rates that resulted in relatively low upper-layer temperatures, below 600 K (621 °F). Fires in compartments smaller than those in nuclear power plants are generally not constant but grow with time, and the upper-layer temperatures frequently exceed flashover levels (about 870 K (1110 °F). Thermal radiation is, of course, a function of the fourth power of the absolute temperature. COMPBRN III results will need to be validated for scenarios during which fire grows with time, or the upper layer reaches a temperature greater than ~650 K (711 °F).

C.2 A PROPOSED TREATMENT FOR FIRE MODELING UNCERTAINTY

The preceding sections of this chapter discuss the specific contributors to the uncertainties in the results of a fire PRA. These sources of uncertainties are identified for those modeling tools and data commonly used in recent fire PRAs, and they may not be applicable to more advanced tools and data which could be used in such analyses. The sources of uncertainties are artificially categorized in two groups—modeling and data uncertainties. "Modeling uncertainties" mainly refers to those sources of uncertainties that stem from commonly used fire propagation models. Error in code predictions for those cases that involve phenomena beyond the applicability of the code assumptions are also treated as additional contributors to uncertainties. This

proposal did not attempt to formally quantify the uncertainties or to explicitly differentiate between uncertainty, variability, and inaccuracy in code prediction.

Identifying the sources of modeling uncertainties in currently available fire propagation computer codes, but not quantifying them, has resulted in a general mistrust in fire code predictions. This is in contrast to the misleading precision of the current fire regulations. It is well accepted in the technical community that the fire combustion, fluid mechanic, and heat transfer phenomena occurring during a fire scenario are quite complex. It is also accepted that the current fire modeling computer codes provide a somewhat simplified picture (in varying degrees) of the phenomena involved. Therefore, it would be quite natural to identify a large number of deficiencies in such codes when applied to a specific fire scenario. Acknowledging the existence of uncertainty is preferable to ignoring it, but there is a danger that such codes will not be utilized because of unresolved uncertainty issues. A deterministic approach to this problem is to limit the utilization of the code to only those fire scenarios or case runs in which the uncertainties are judged to be small, therefore justifying its applicability. There are two fundamental flaws with such an approach: (1) asserting whether the uncertainties are small or not implies that they are quantified and (2) identifying a comprehensive set of configurations and parameters for which a computer code, comprising many models and submodels, could be effectively used may not be possible without severely limiting the application domain of the code.

Contrary to the deterministic approach, the probabilistic approach requires that modeling uncertainties be quantified in a formal manner for each case run and the decision be left to the user in light of variabilities of the results predicted. However, the probabilistic approach cannot be utilized unless we proceed beyond the current haphazard, qualitative treatment of the modeling uncertainty. Although the rationale for the probabilistic approach has long been accepted, there is little or no consensus on the methodologies to be employed.

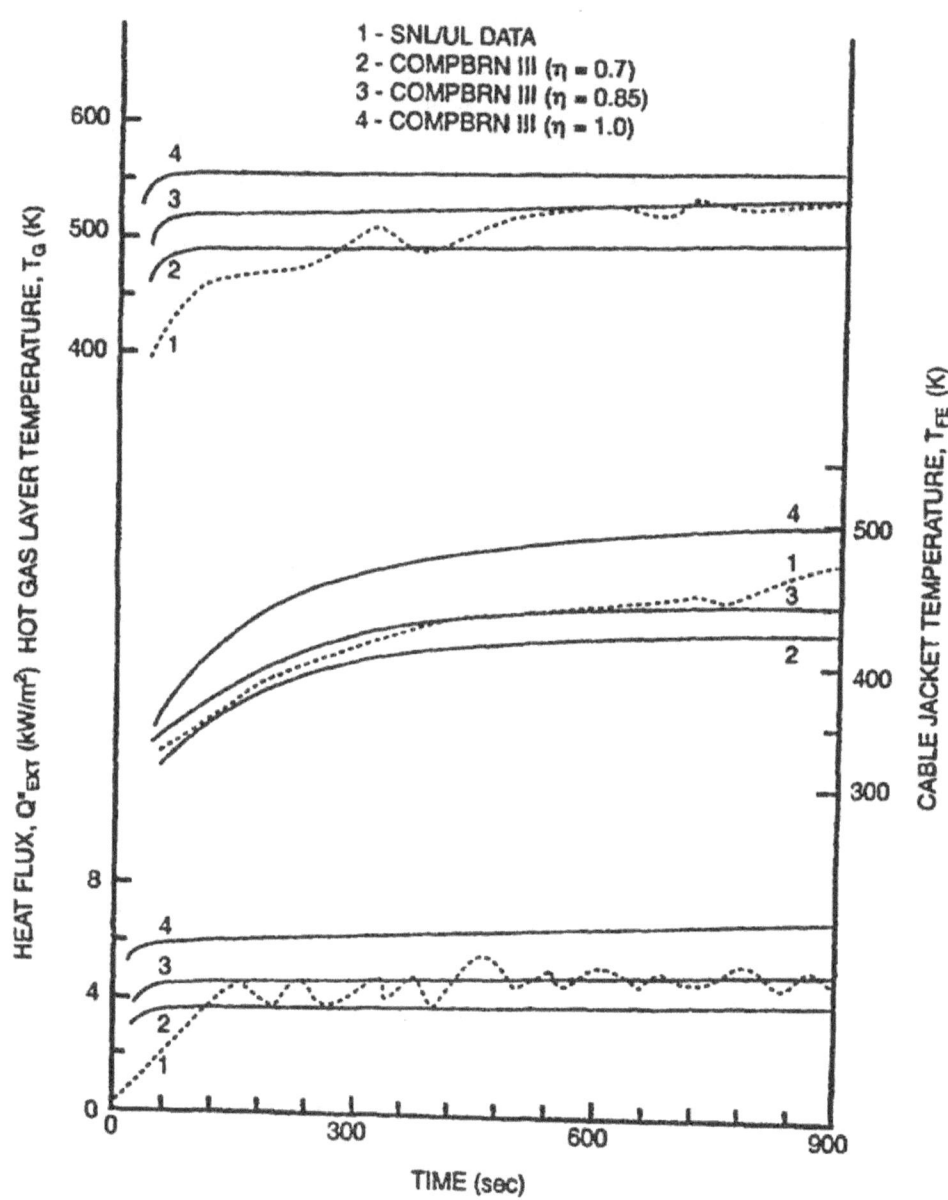

Reproduced From NUREG/CR-3192

Figure C.4
Simulation of SNL/UL Experiment 1

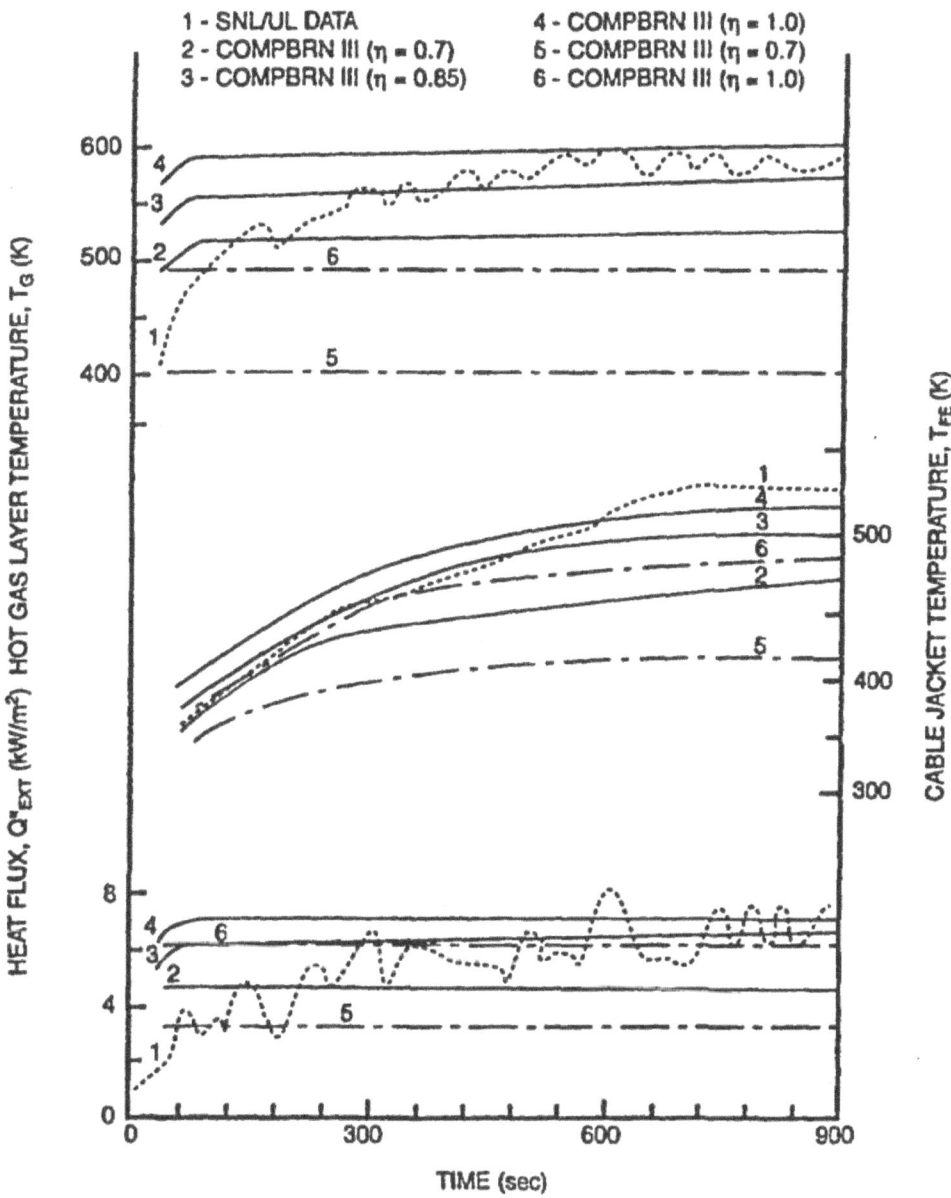

1 - SNL/UL DATA   4 - COMPBRN III (η = 1.0)
2 - COMPBRN III (η = 0.7)   5 - COMPBRN III (η = 0.7)
3 - COMPBRN III (η = 0.85)   6 - COMPBRN III (η = 1.0)

Reproduced From NUREG/CR-3192

Figure C.5
Simulation of SNL/UL Experiment 2

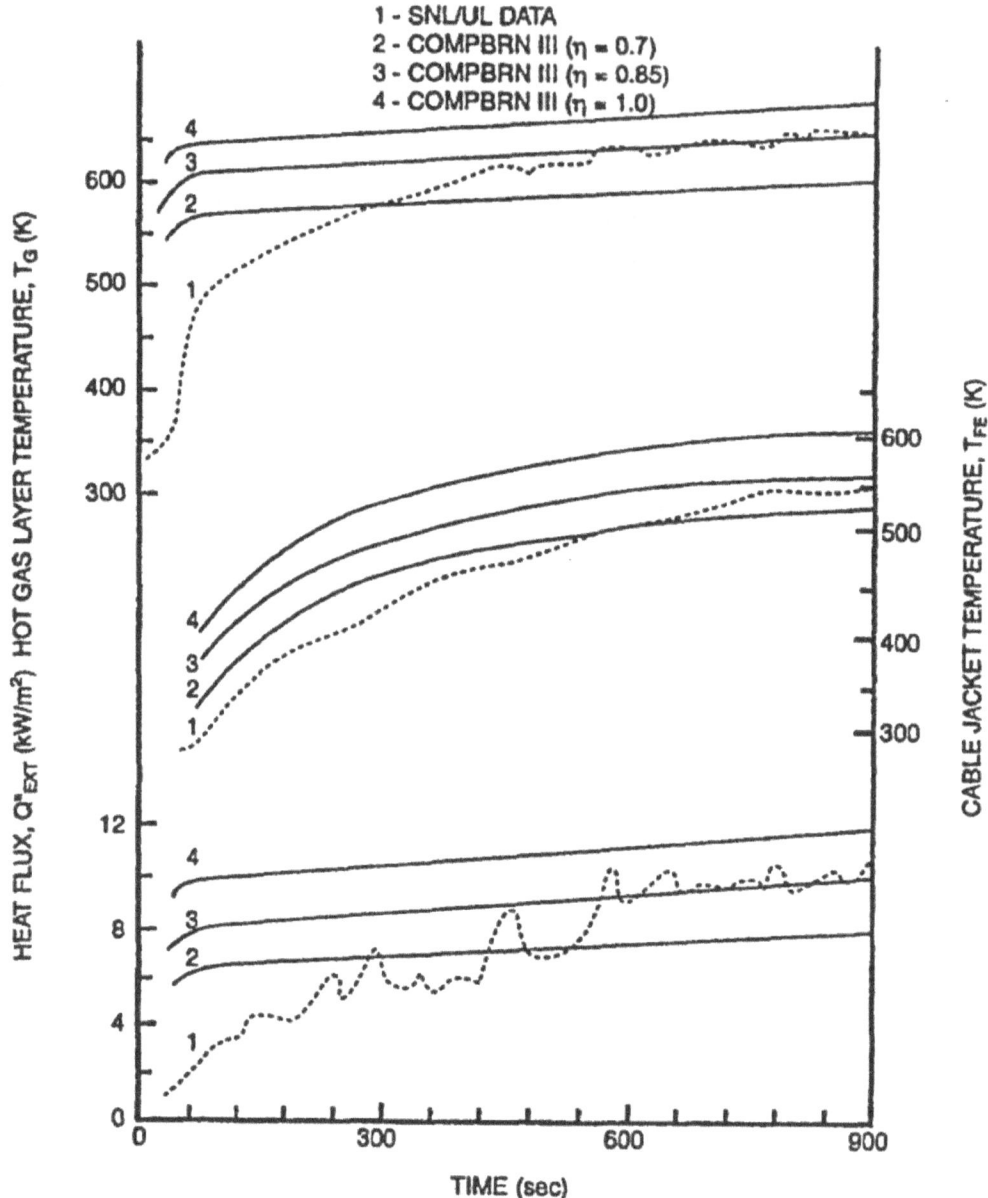

Reproduced From NUREG/CR-3192

Figure C.6
Simulation of SNL/UL Experiment 3

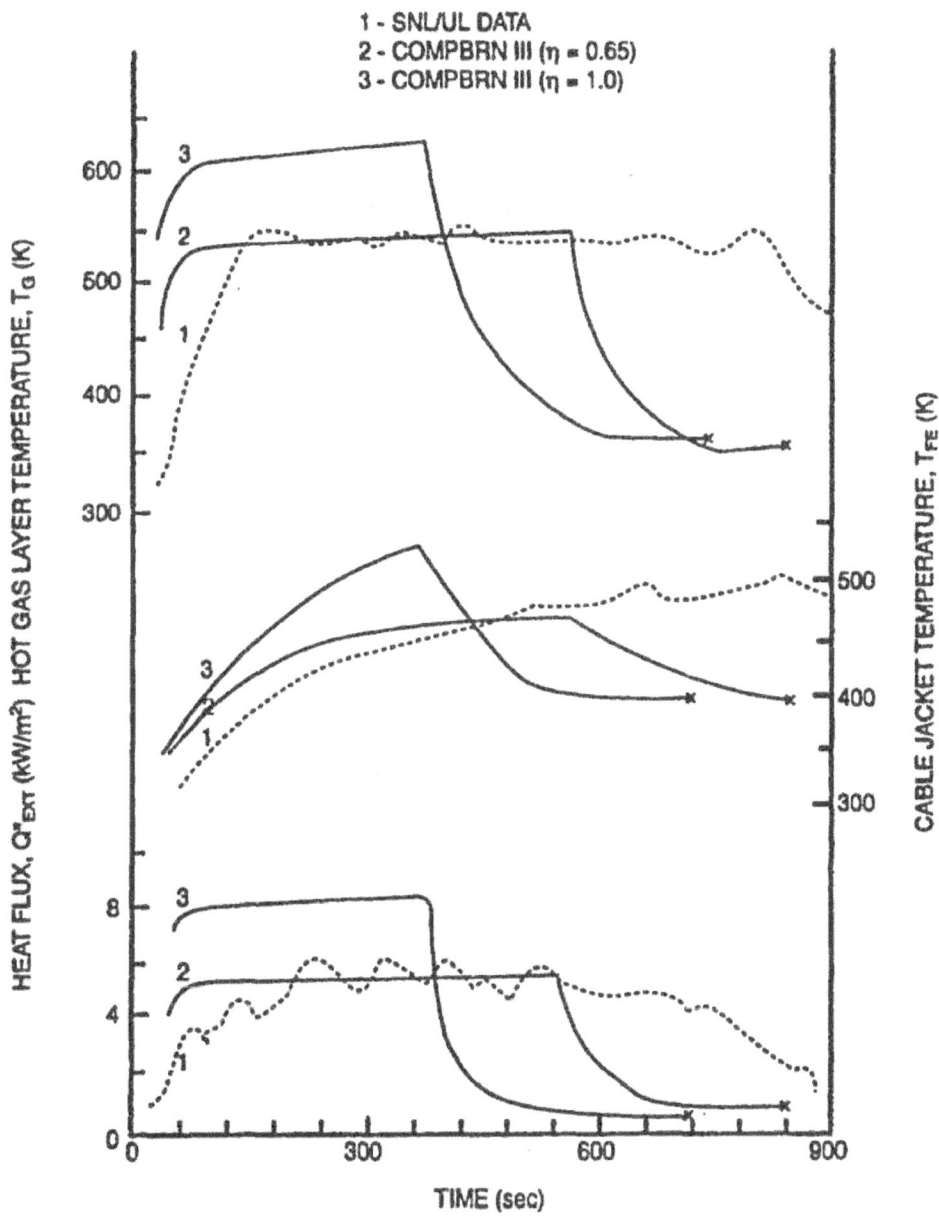

1 - SNL/UL DATA
2 - COMPBRN III (η = 0.65)
3 - COMPBRN III (η = 1.0)

Reproduced From NUREG/CR-3192

Figure C.7
Simulation of SNL/UL Experiment 4

This section proposes an approach for evaluating modeling uncertainties. In this approach, modeling uncertainties are estimated at the lowest level of modeling at which experimental data are available and are propagated through the various interconnected modules. This section also contains a perspective on modeling vs. data uncertainties as it pertains to fire propagation modeling. It also discusses some approaches for formal evaluation of such uncertainties and, more importantly, practical uses of code predictions in a decisionmaking process.

C.2.1 Model: A Simple Definition

A model is a mathematical description of a theory, sometimes under certain restrictions or assumptions, which can explain a set of experimental observations and predict the results of a similar experiment not yet observed. Models have a specific objective which defines what the model will predict, and a minimum set of restrictions that are required for the prediction to be valid.* Several models can be developed for a given theory depending on the objectives chosen for the model and the restrictions imposed on the theory. Therefore, some models can have wider applications than others if they contain less restrictive assumptions and simplifications.

C.2.2 Model vs. Parameter Uncertainty

A model has a mathematical form (shown by an operator $L(.)$) containing input variables** represented by vector X, an output variable shown by vector Y, and parameter variables (a part of data uncertainty) given by vector α : $Y = L (X, \alpha)$.

The theory upon which a model is based typically

---

* As an example, a model can be constructed on the basis of the conservation of the momentum in a buoyant fluid with an objective to predict the mean velocity assuming a top hat velocity profile for the fluid as typically used for semi-empirical formulation of the fire plume.

** More generally referred to here as influencing variables.

identifies the relationship of the influencing (input) variable X on the output variable Y. The first question of concern is the question of completeness, i.e., are all the influencing variables accounted for in the model? In almost all of the models used for engineering calculations, some influencing parameters are not accounted for. The modeler usually neglects some contributors to facilitate obtaining a solution to the model. This greatly influences the domain of model applicability. However, the modeler assures that, for the domain of the applicability, the effects of such approximations are small and can be accounted for by the parameter's uncertainties. It should be noted that to neglect some contributors, one may not need to know the universal form of the operator L. For example, one may choose to neglect all influencing variables of a short-term transient if the objective of the model is to evaluate the long-term transient response. Here, the modeler neglects the effect of the short-term transients and the associated influencing variable only on the basis of the objective of the model, but not necessarily on the basis of universal form of the operator L. In reality, the universal form of operator L is only known for a few fundamental theories.

Given the objective and the influencing variables (Y and X), and a set of experimental measurements of X and Y, the analyst completes the model by selecting appropriate forms for L and proper statistical estimates for α . There is obviously an interplay between the form L and the parameter vector set α . Generally for any given L, statistical estimates for α and their associated uncertainties can be obtained. For this reason, many believe that the uncertainty in a model structure L can always be represented by parameter uncertainties. Using the same argument, the parameter uncertainties could also be represented by the model uncertainty. This interplay between the parameter and model uncertainty could allow the aggregation of both types of uncertainties in a form of either the parameter or model uncertainty for a domain for which experimental data are observed. However, if the model is used for its primary objective, which is the prediction of the output variables for a domain of input variables X for which no experimental data arc available, the form of the operator L can significantly change the prediction or the extrapolation of the experimental results. In this case the modeling uncertainties (represented by different forms of the operator L, e.g., L_j) would not be captured by the parameter uncertainties. Generally, parameter uncertainties are more important for interpolations within the experimental domain (since parameter uncertainties are estimated and adjusted to reflect the variations in the existing experimental results), whereas modeling uncertainty would be more important in extrapolation outside the experimental results. It would be also a matter of convenience and consistency to differentiate between the model and parameter uncertainties for both interpolation and extrapolation of the experimental results.

The operator L can take the form of a function, a set of partial differential equations, integral equations, or a combination thereof. When operator L describes a process with initial and boundary conditions, another source of uncertainty enters the results. In most modeling applications to PRAs, the initial and boundary conditions are uncertain. An example of the initial condition could be the ambient temperature in a room when a fire starts. The room temperature may vary significantly throughout the year (seasonal variation). Therefore, this initial condition would be uncertain since fire could occur at any time of year. Another example related to fire PRA would be the initial size of the transient fire as the initial condition for fire propagation modeling. The uncertainties associated with initial and boundary conditions in PRA applications are typically irreducible uncertainties (or what is usually referred to as "variability"). That is, collecting more detailed information from the fire events in nuclear power plants may not necessarily reduce the uncertainties in the initial fire size, but may provide better estimates on its uncertainty parameters. The uncertainties associated with initial and boundary conditions are currently being treated as a part of data uncertainty in some fire computer codes. In other fire analyses, the uncertainties of the initial fire size are not explicitly accounted for. In these analyses, a

maximum credible initial fire size is typically selected to show the capability of the fire protection systems. This is similar to use of models in design applications, where the initial and boundary conditions are typically assumed, rather than formally addressed.

The last source of uncertainties in modeling prediction could result from the use of numerical algorithm and nodalization (or discretization) for solving the model. Such numerical inaccuracies could be generally reduced. These inaccuracies are not expected to contribute significantly to the uncertainties in model prediction for most of the commonly accepted codes.

Generally, the parameter uncertainty, the uncertainties in initial and boundary conditions, and the inaccuracy of numerical algorithm are referred to as "data uncertainty." The modeling uncertainty, therefore, deals only with the various acceptable forms of the operator L.

C.2.3 How To Quantify Modeling Uncertainty

Formal methodologies are currently available to evaluate the parameter uncertainties and the variability in initial and boundary conditions (referred to as "data uncertainty"). However research results and formal methods for evaluating modeling uncertainty are sparse and application results do not exist. A framework recently proposed by Apostolakis (see NUREG/CP-0138) is commonly referred to as the $P\{M_i\}$ framework. In this framework model, uncertainty is measured by a probability distribution over a given set of model operators L_i. As discussed in the workshop (see NUREG/CP-0138), a number of difficulties that can arise from such an interpretation of modeling uncertainty and its subsequent quantification. We therefore propose, albeit informally, another approach for interpretation and quantification of the model uncertainty. In this approach, model uncertainties are decomposed and treated at the basic module levels of a computer code. The code is typically considered as a number of modules and submodules that are integrated by proper assignments of outputs of one or more submodules to the inputs of other modules. Various sources of uncertainties, including the modeling uncertainty, then are propagated from one module to another to arrive at the final uncertainties of the code predictions. To do this, we define a new random variable γ to represent the modeling uncertainty for each module. That is

$$Y = L(X, \alpha) + \gamma \qquad (C-1)$$

The probability distribution for random variable γ could be conditional on X, or on some distance measure of X. An example of a distance measure could be a Euclidean distance of X from those values of X that are observed by experiments (Euclidean distance of a point from a cluster of experimental points). Other measures of distance could be envisioned, especially those that are normalized. An alternate form of Equation C–1 may be written by expressing γ by a residue-random-variable R_X and the expectation of the operator L over all values of α. That is

$$Y = L(X, \alpha) + R_X\{E_\alpha[L(X,\alpha)]\} \qquad (C-2)$$

Taking expectations over α and the residue variable R_X from both sides of Equation C–2 for a given value of X results in

$$E_{\alpha,R}(Y) = E_\alpha[L(X,\alpha)] + E(R_X)\{E_\alpha[L(X,\alpha)]\}.$$

Now if we consider the existence of a model L^* for which the $E(R_X)$ is zero (sometimes referred to as a best-estimate model), then for any model L_j with the associated residue function R_j, the following expression holds:

$$E(R_j) = (L^* - L_j)/L_j \qquad (C-3)$$

Equation C–3 basically describes the close relationship of the expectation of R_j with the degree of bias embedded in the selected model for the module under study. For a conservative model, this is referred to as a safety factor. The degree of bias in a model simply results from a conservative assumption. As an example, heat losses may only be considered through the ceiling and not for the side walls in a fire model. Obviously, such an assumption introduces conservatism which will eventually result in overestimations of the temperature and the thickness of the hot layer. The variance of R_j is similarly related to the mathematical approximation of the underlying physical phenomenon. As an example, the solution of a heat conduction model for the temperature profile in a finite slab could be approximated by neglecting the second-order and higher order terms in the appropriate expansions. Such an approximate model could be used to obtain an estimate for slab heatup calculations (change of mean temperature vs. time). Therefore, a simpler model may underestimate the variance of the results compared to a more accurate model if only parameter uncertainties are considered. The variance of R_j, therefore, should be larger in a less rigorous model to compensate for this underestimation of variance.

Engineering and scientific considerations can be used to explore the characteristics of R_j for the domain of application for each module based on the experimental information that is available. Treatment of modeling uncertainty at the module level also facilitates the use of both the available experimental results and the results from more rigorous modular codes for the determination and calibration of the characteristics of R_j. These modular uncertainty characteris-tics then can be propagated through an integrated code to arrive at a distribution for the plausible code predictions, accounting for all known uncertainties.

This approach relies heavily on the results of experiments for each module rather than on the results of integral tests. The results of integral tests are used mainly for validation to show that the test results are within the uncertainty ranges of the code predictions and that the code predictions are unbiased. On the contrary, the experimental results for each module are directly used to estimate both the modeling and the parameter uncertainties. Each module can be generally categorized into one of the three groups that follow:

(1) *Physically based module*: This is a modeling module for which the underlying physics is well understood, and the uncertainties

mainly stem from model simplification, numerical algorithms, and the uncertainties associated with initial and boundary conditions.

(2) *Semi-empirical module:* This is a modeling module for which the influencing variable can be identified and a qualitative relationship (but not the exact equation) between input and output variables can be established, e.g., pressure and temperature are monotonic. Here the experimental results can be used to determine the most appropriate functional form and its associated uncertainties.

(3) *Empirical module:* This is a modeling module for which the level of physical understanding is poor and consensus is not established among experts. The experimental results are typically sparse in this category. Therefore, the experts may propose different models—each with different implications. The proposed $P\{M_i\}$ framework discussed earlier would be most suited for this category. Sparse experimental data may be used to appropriately weigh different expert judgment. Another approach that is more consistent with our earlier framework is to take the average of all the functional forms proposed by the experts, and to show the variation among experts by the distribution of the residue variable R_j. The prior distribution for R_j obtained in this manner then could be used in a Bayesian updating routine to arrive at the posterior uncertainty distribution for each module when sparse test results are available.

A computer code may contain several modules in each category defined above, depending on the state-of-the-art knowledge for the phenomena represented by the modules. Some thermo-hydraulic codes may contain mostly category 1 modules, whereas some severe-accident codes may contain several modules in categories 2 and 3. A fire propagation computer code is expected to have modules that belong to each of the preceding three categories. For example, the plume module would be category 1, cable damage

criteria would be category 2, and burning of the cables and the associated heat-release rate may be considered as the category 3 module at the present time (see discussion in Sections C.1 and C.2).

C.2.4 Decisionmaking Under Uncertainty

When a model is used to predict the outcome of a scenario of interest, a decision can be made to either accept or reject the final outcome. As an example, a computer code may be used to estimate the peak cladding temperature for a given scenario of interest and to compare it with an acceptable criterion (i.e., 2200 °F). If the results of the code indicate that the cladding temperature never exceeds the criterion, then a decision may be made that the plant can safely respond to that transient. However, to arrive at that decision the analyst typically evaluates the following considerations:

(1) *Qualitative considerations:* Here the decisionmaker evaluates the technical details of the various modules within the model or the computer code. The focus is to identify the scope of the modeling and how it compares with the phenomenon of interest in the scenario. The analyst checks that the methods used are well documented, their limitations are well known, and so forth. This step basically establishes the credibility of the model and its applicability.

(2) *Quantitative considerations:* Here the decisionmaker evaluates the summary results of the computer code. This evaluation is typically done by model verification and validation. Verification and validation typically involve comparing the code summary results with the results of more sophisticated codes or experiments. In most cases, this type of evaluation results in code or model calibration. Every computer code has a set of tuning parameters that can be adjusted to result in a closer estimate of "reality." Here the word reality is enclosed by quotation marks to indicate that reality refers to results that are more acceptable to the analyst (e.g., from a more sophisticated code) and that are expected to be a more

accurate representation of the true outcome. Other ways of calibration involve the introduction of bias factors reflecting the degree of the conservatism or unconservatism in the code models. Calibration of the models and the computer codes are application specific. It is generally expected that, for a set of applications or scenarios which involve similar initial and boundary conditions as well as comparable ranges of influencing variables, the calibrating and biases factors remain unchanged. Such calibrations typically result in an unbiased or a best-estimate code.

(3) *Probabilistic considerations*: Here the decisionmaker is concerned with the final decision for the specific scenario analyzed, on the basis of the summary results generated by the code. The decisionmaker (perhaps the regulator) is aware that the results generated by the code are accurate within a certain error bound. In our earlier example, if the code predicts a peak cladding temperature of 2150 °F, compared to the 2200 °F criterion, the regulator may decide that the criterion is not met. (The regulator in a sense believes that the code prediction is not accurate within 50 °F.) This problem is traditionally treated informally in an ad hoc manner with the use of safety factors. The regulator commonly uses either a conservative criterion (e.g., 2000 °F instead of 2200 °F) or a conservative analysis with the use of a multiplier. In some cases, both are used. Probabilistic analysis, on the contrary, is a formal methodology that quantifies the uncertainties from both the model and data, and it allows an estimation of the probability that a decision is true (confidence level). Conse-quences of decision alternatives may be evaluated and compared formally, and the final decision can be optimized on the basis of a given cost function, if so desired. In most cases, a regulator is interested in the outcome that has a high level of confidence (95 percent or more). For this reason, uncertainty evaluation should become an integral part of the decision process.

C.2.5 Considerations of Uncertainties in Fire Modeling

A fire modeling computer code for use in fire risk assessment in nuclear power plants should provide the following minimum information:

*   time of activation of fire detectors

*   time of activation of fire suppression systems

*   time of damage of critical targets and equipment

*   time of flashover

*   time of barrier failure and propagation to other rooms

*   time of fire burnout

These objectives are met by predicting the local concentration of combustible products, humidity, and other thermohydraulic characteristics, such as gas temperature and velocity. A fire computer code is typically written in modules or submodels that are integrated by proper assignments of outputs of one submodel to the inputs of other modules. Various sources of uncertainty, therefore, are propagated from one module to another in an integrated code. The modeling uncertainties for each module, therefore, should cover large ranges of the influencing variables. As noted earlier, the dependence of modeling uncertainty (i.e., mean and variance) with the values of the influencing variable should be accounted for with some kind of normalized distance measures. Various sources of experimental and analytical data are typically available at the sub-model level to estimate the parameters of the modeling uncertainty distribution. The formal evaluation of modeling uncertainty is both costly and time consuming. Therefore, the analyst should focus on the major sources of this uncertainty. Section C.2 illustrates a process to characterize the major sources of modeling uncertainties in current fire computer codes where the study should focus. It is sometimes more beneficial to substitute more comprehensive models (if available and practical) for deficient code modules, rather than formally estimating the resulting modeling uncertainties. As an example, for a fire computer code, it might be more prudent to model the effect of the oxygen availability for predicting fire heat-release rate, rather than treating it as a source of uncertainty. Both subjective evaluation and sensitivity runs are helpful to decompose and prioritize the sources of uncertainties and to identify those areas of modeling that can be easily refined.

In addition to performing uncertainty analyses at the submodel level, some authors have recommended evaluating modeling uncertainty at the code level when integrated test results are available. Methods such as the use of mixture distribution were recommended by the Nuclear Safety Analysis Center in NSAC 181. We feel that integral test are important for code verification and for understanding the interactions among various phenomena involved in the scenario. The results of integral tests, when decomposed to different phenomena and submodels, could be used in the approach discussed here to estimate both the modeling and the data uncertainty.

C.3 SUMMARY

Various uncertainty issues associated with risk-informed and performance-based approaches specific to fire protection requirements are discussed in this chapter including those with fire modeling. Many fire protection requirements may be evaluated without the need for fire modeling (e.g., surveillance issues and system issues). For these cases, the issue of uncertainty can be formally addressed and incorporated in the decisionmaking process. In other cases in which evaluation of the requirement necessitates the use of fire modeling, the portion of fire modeling that

predicts the fire heat-release rate was differentiated from the portion that predicts the thermal environment. Larger uncertainty ranges are associated with the predicted heat-release rate than with the thermal environment. The heat-release rate is the driving force for the plume mass flow rate, the ceiling jet temperature, and finally, the hot layer temperature that is driven by energy balance. The fire heat-release rate is dependent on the initial fire size, the growth of fire by propagation and ignition of additional combustibles, and the heat-release rate from these additional combustibles. Current computer codes are judged to perform sound analyses of thermal environments, and some may carry formal uncertainty evaluation. On the other hand, current codes either do not model the source fire heat-release rate or the treatment is valid only under certain conditions. In any case, the heat-release rate of the fire source, knowing the current state of the art, may be best estimated conservatively by using simplified engineering evaluation, subjective judgment, and extrapolation of actual fire events or fire tests.

Some definitions for modeling and data uncertainties are proposed in Section C.3. Several sources of data uncertainties, i.e., parameter uncertainty and uncertainty of initial and boundary conditionsm are identified. The current treatment of data uncertainties is summarized and different sources of modeling uncertainties resulting from assumptions, approximations, simplifications, and numerical algorithms are discussed. An approach is proposed on the basis of decomposition of uncertainties to the most basic level of modeling and aggregation of the uncertainties using the current uncertainty propagation techniques. A process for decisionmaking under both modeling and data uncertainty is also presented.

Appendix D

APPLICATIONS OF RISK-INFORMED, PERFORMANCE-BASED METHODS

CONTENTS

APPENDIX D
APPLICATIONS OF RISK-INFORMED, PERFORMANCE-BASED METHODS

D.1 FORMAL UNCERTAINTY EVALUATION FOR ANALYSES FOR OPTIMIZING TEST DURATION FOR APPENDIX R EMERGENCY LIGHTING

The case study presented in Section 6.2.1.2.2 shows that reducing the duration of annual testing from 8 to 5 hours could reduce the number of battery replacements, while at the same time, their reliabilities would only marginally change. The analysis assumed that the actual battery rating is normally distributed and the parameters for the normal distributions were subjectively assigned using the available engineering information and test data (see Equations 6–2 and 6–3). Formal uncertainty analysis was not performed since some of the engineering data were qualitative and not amenable to formal quantative uncertainty evaluation. However, it is felt that it would be important to demonstrate the uncertainty evaluation methodologies in this section by assigning quantitative values to those measures where only qualitative information is currently available.

The analyses for this case study consists of three modules:

(1) semi-empirical models to determine the failure probability of an emergency light when demanded for 8 hours of continuous operation, given several previous full discharges

(2) reliability replacement models to determine the expected number of full discharge tests that each unit (out of a population of emergency lights) could have experienced, accounting for replacement after failure

(3) reliability integration models to determine the failure probability for the minimum number of emergency lights required for a successful demand

Formal and defensible uncertainty evaluation for this case study would require the availability of specific test data to estimate the parameters of the models. Such parameters are not currently available even though they could be obtained either from the manufacturer or by a set of tests, as discussed later. Regardless of the availability of the specific data, an uncertainty evaluation could still be performed using subjective estimates on the uncertainty range of these parameters. For this demonstration, the following discussion will concentrate on the first modeling module. The modeling parameters and modeling assumptions that are subject to variation are identified, followed by a discussion of the sources of uncertainties and the specific test data that can be used to estimate the expected variations.

The actual rating of a rechargeable battery can be described by the following expression:

Actual Rating = Manufacturer's Rating × Margin Factor × Effect of Previous Discharges × Effect of Temperature

Manufacturer's Rating

The following assumptions are made for the manufacturer's rating:

- Modeling Assumption: Coefficient

- Parameter Uncertainty: None (This is actually the name-plate rating.)

Margin Factor

The following assumptions are made for the margin factor:

- Modeling Assumption: Coefficient $= 1 + \beta$

- Parameter Uncertainty: The uncertainty would represent the variation in margin for different manufacturers and for the same manufacturer, but different manufacturing batches. In most cases, the manufacturer or the batch records may not be available and both sources of variability should be included in the analysis. Testing different types of batteries to failure would provide the necessary information for estimating this factor. For this analysis, the parameter β is subjectively assumed to be lognormally distributed with the mean of 0.15 and an error factor of 3.

Effect of Previous Discharges

The number of full discharges is expected to reduce the battery's capabilities. Generally, batteries that are fully discharged (based on their manufacturer's rating) more than 20 times are not considered reliable. The relationship between the number of discharges and the effect on battery rating is not clear. The relationship could be presented through a concave, linear, or convex curve. A linear model was used in the previous point estimate calculations.

- Modeling Assumption: The effect is shown by a family of curves based on the value of a: between 0.2 and 1.8, as follows:

$$(1 - N_d F_d)^a$$

where

N_d = the number of demands
F_d = the discharge coefficient.

- Parameter Uncertainty: The uncertainties associated with the above parameters represent the design and manufacturing variability in the useful life of a battery in terms of the number of discharges in a controlled temperature environment. The following subjective uncertainty distributions are assigned to the above parameters:

  a: is uniformly distributed between 0.2 and 1.8.
  F_d: is lognormally distributed, with the mean of 0.05 and an error factor of 2.
  N_d: is the number of discharges and is not a stochastic parameter.

Effect of Temperature

The effect of temperature on the rating is estimated on the basis of empirical cell-size correction factors tabulated in ANS/IEEE 485-1983 (see Table 6.2). The effect of temperature is the reciprocal of the correction factor shown in this table.

A quadratic curve fit to the data in the table resulted in the following dependence of the correction factor on the temperature: Correction Factor = $1.4 - 0.02157\,t + 0.0002181\,t^2$.

Mean square error is less than 3 percent for each coefficient of the quadratic function, i.e., this function excellently approximates the data. The actual temperature in the room at the time of demand is a stochastic variable which varies from one room to another, and depends on the heating, ventilation, and air conditioning (HVAC) system and temperature control at the room, the seasonal variation of the outside temperature, and the location of the plant and its associated climates. This type of information is easily obtainable for a given room. For this case study the following assumptions are made:

- Modeling Assumption: Empirically based models

- Parameter Uncertainty:The equivalent temperature at the time of demand is assumed to be normally distributed with the mean of 18 and standard deviation of 8 °C.

Using a Monte Carlo simulation and the developed model, the uncertainty in the battery rating for different numbers of tests is evaluated. Also, a probability of failure of an 8-hour-rated battery-operated emergency light with 8 hours of mission time is calculated.

The present model contains four uncertain parameters: parameter β in the margin factor, parameters F_d and a, defining effects of previous discharges, and temperature t variations. Although, from a modeling point of view, uncertainties in these four variables are treated similarly, they are quite different from a physical point of view. We can categorize these uncertainties in two groups:

- *irreducible uncertainties* or *variabilities* which cannot be eliminated (variance cannot be reduced to zero) by additional data, experiments, and tests and

- *reducible uncertainties* which can be eliminated if sufficient tests are carried out, i.e., these random variables can be reduced to deterministic values if a sufficient amount of data is available.

Evidently, temperature variations should be classified as irreducible uncertainties, unless a temperature control system is installed. Also, variability in parameter F_d in effects of previous discharges cannot be reduced to zero because of variabilities in physical processes and in manufacturing parameters (e.g., quality of materials, dimensions). An uncertainty in the modeling parameter a can be reduced to zero if there are sufficient failure statistics, i.e., we can find the best modeling parameter a. Parameter β in the margin factor can be treated in both ways: variability of this parameter cannot be reduced to zero; nevertheless, we can find some low bound for this parameter, like a 5-percent quantile, and use it as a deterministic value in the reliability evaluations.

The Monte Carlo simulation of the battery rating uses the model presented in the previous section. The simulation code was implemented with the MATHEMATICA package. The number of tests (8-hour discharges) were varied $N_d = 0, 1, ... , 20$, and different statistical characteristics of the rating (mean value, 5th, 25th, 75th, and 95th percentiles) were evaluated. To address the impact of uncertainties, four cases were analyzed:

Case 1: Uncertainties in all random parameters

Case 2: Uncertainties in the parameters F_d and a (which define the effects of previous discharges) and in temperature t

Case 3: Uncertainties in the parameter F_d and in temperature t

Case 4: Uncertainties in temperature t

Uncertainty is removed in the margin factor β, then in the modeling parameter a, and then in the parameter F_d. Figures D.1 through D.4 are graphs representing Cases 1, 2, 3, and 4, respectively. For instance, Figure D.3 shows that with 11 demands, the mean average rating equals 4 hours. There is an irreducible uncertainty in this value: with 5-percent probability, the actual rating equals zero. Comparing Figures D.3 and D.4, we see that the random coefficient F_d contributes significantly to the uncertainty in the actual rating. When there are more than 10 tests, and there is an uncertainty in F_d, there is at least a 5-percent chance that the actual rating equals zero. However, without uncertainty in this random value, the 5-percent quantile does not reach the zero value.

Finally, the probability of failure of a battery-operated light with an 8-hour mission time (8-hour-rated battery) is evaluated. Figure D.5 presents this probability as a function of the number of tests; the probability ranges from 0.4 to 1. It equals 0.4 for a new battery and 1.0 if the number of tests exceeds 10. These values are slightly higher than the battery failure rates estimated earlier without formal treatment of uncertainty. The present study assumes a large temperature variation in the room (rather than the fixed temperature of 77 °F assumed earlier) and generally uses lognormal distributions (which result in more conservative estimates of the mean). With such major differences between the two approaches, it is quite encouraging that the results are so close. Figure D.5 does not present uncertainty bounds because the model does not include uncertainties in the means and variances of the random values. The current assumption in the model is that we know the distribution parameters exactly. However, including uncertainties in parameters of the uncertainty distributions allows us to evaluate uncertainties in failure probabilities. Using conservative estimates for distributions, we can obtain conservative estimates for the failure probabilities. The conservative estimates are preferable for most practical applications with highly reliable components and systems.

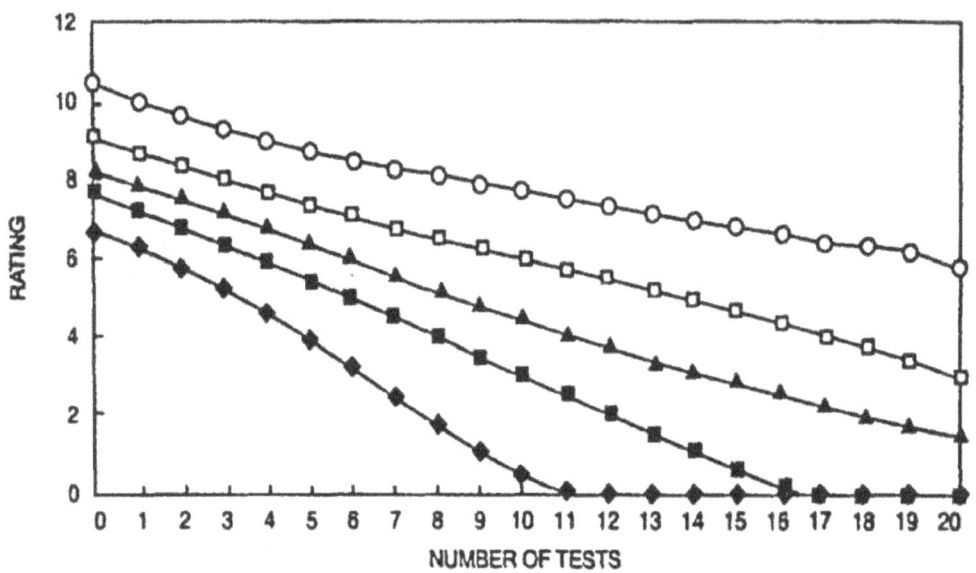

Figure D.1
Uncertainties in All Random Parameters

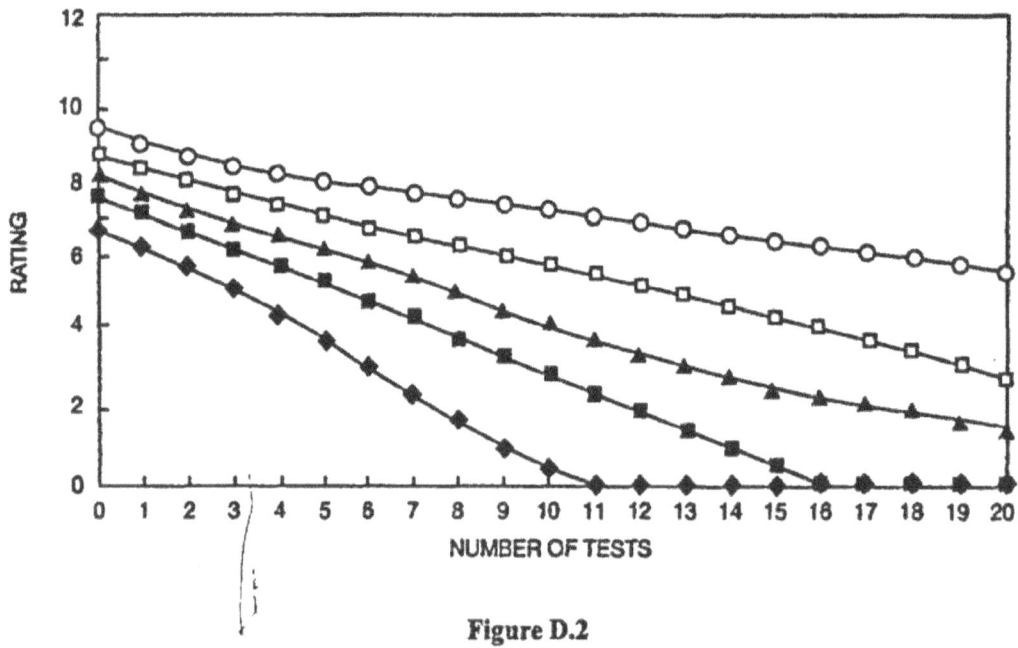

Figure D.2
Uncertainties in Parameters F_d, a, and in Temperature t

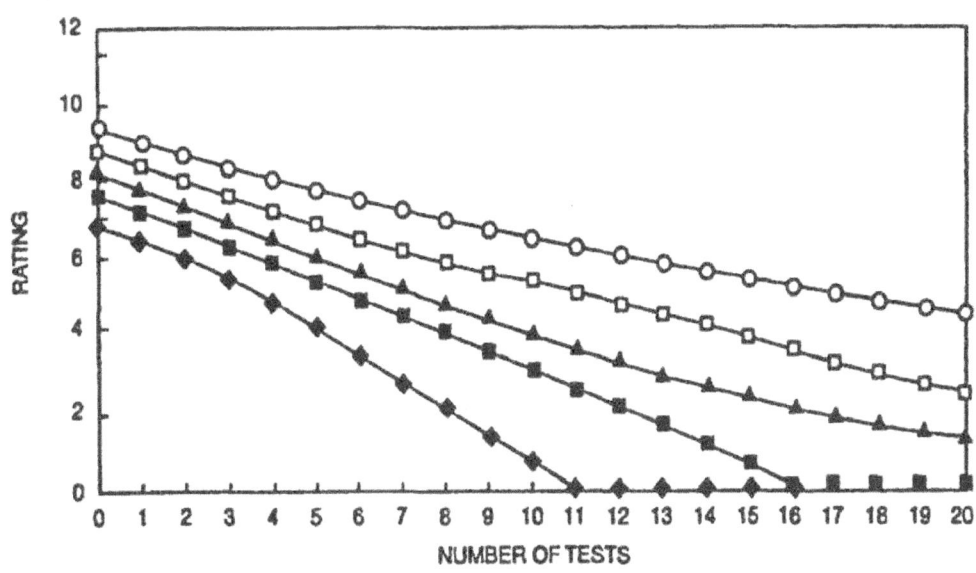

Figure D.3
Uncertainties in Parameter F_d and in Temperature t

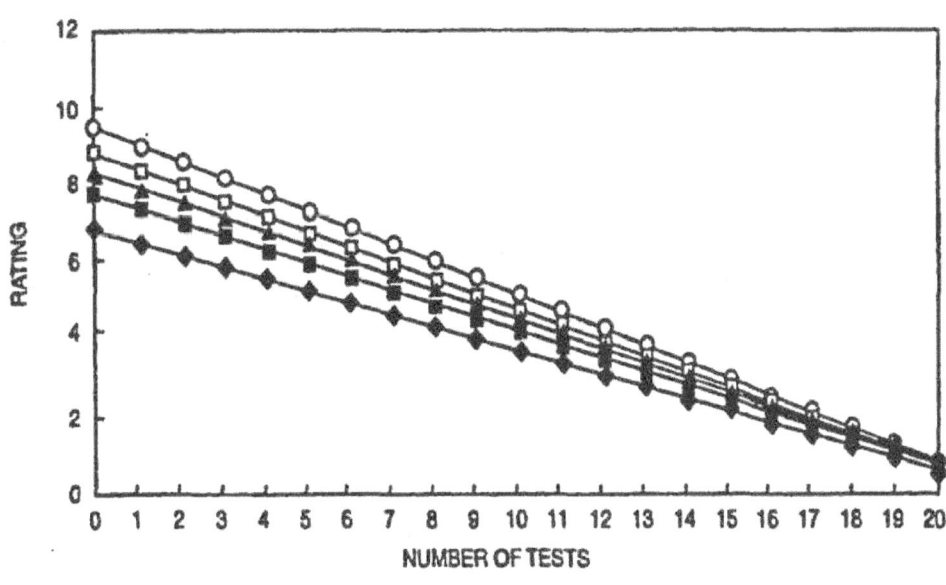

Figure D.4
Uncertainties in Temperature t

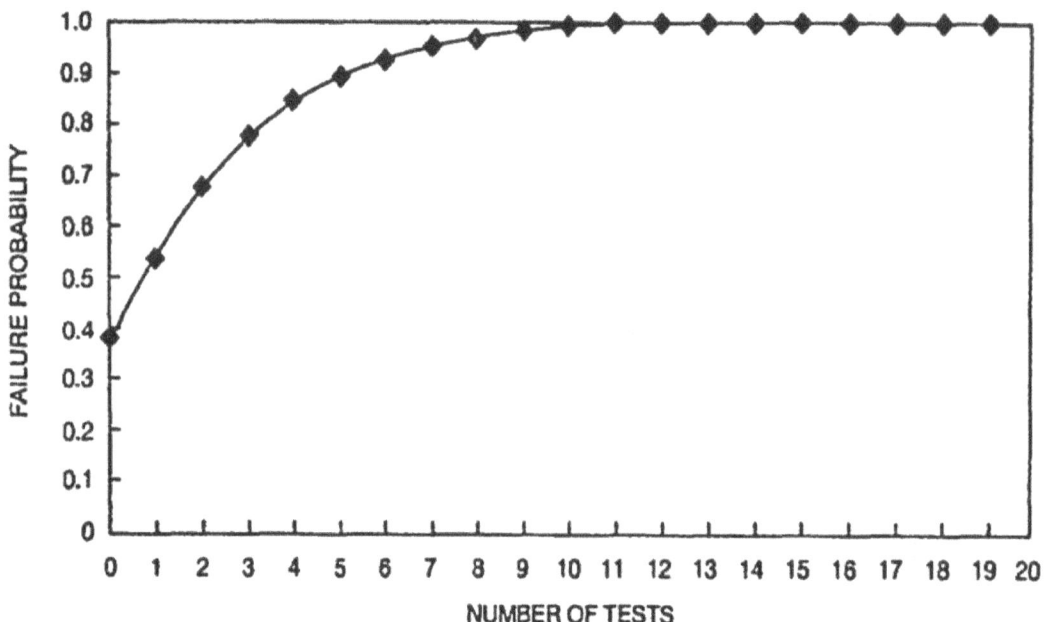

Figure D.5
Failure Probability for Battery-Operated Emergency Lights
(8-Hour Mission Time, 8-Hour-Rated Battery)

D.2 FIRE MODELS AND COMPUTER CODES BASED ON ZONE MODELS—ANALYSIS OF SAFE SEPARATION DISTANCE

The fire protection requirement of safe-shutdown capability is contained in Section III.G of Appendix R. Section III.G requires that one train of systems necessary to achieve and maintain hot shutdown conditions from either the control room or emergency control station(s) be free of fire damage. This requirement is met using one of the following strategies:

(1) Separation of cables, equipment, and associated non-safety circuits of redundant safe-shutdown trains by a fire barrier having a 3-hour rating.

(2) Separation of cables, equipment, and associated non-safety circuits of redundant safe-shutdown trains by a horizontal distance of more than 6.1 m (20 ft) with no intervening combustible materials or fire

hazards. In addition, fire detectors and an automatic fire suppression system should be installed.

(3) Enclosure of cable, equipment, and associated non-safety circuits of one redundant safe-shutdown train in a fire barrier having a 1-hour rating. In addition, fire detectors and an automatic fire suppression system should be installed.

Finally, if none of the items in (1) through (3) are complied with, alternative or dedicated shutdown capability independent of the fire area under consideration should be provided. At many two-unit sites, cross-connection between safe-shutdown systems is considered as an alternative shutdown capability. The time required for manual alignment of the cross-connections has been a major issue during mid-loop operation in PWRs.

The purpose of presenting this case study is to evaluate technical methods available to examine the risk significance of the 20-ft horizontal separation criterion and for alternative performance-based approaches.

D.2.1 Importance of the Case Study

As discussed briefly in Chapter 3, several licensees have selected strategy 2, but have requested exemptions from the requirements associated with 6.1-m (20-ft) separation or areawide automatic suppression. In almost all cases, some combination of low combustible loading, a high compartment ceiling, or negligible intervening combustible materials is used as justification.

Several exemption requests were reviewed. The following two exemption requests gave specific cost estimates for justifying the burden to the utility if the exemption was not approved:

(1) Cable rerouting and an alternative power source for either high pressure coolant injection or reactor core isolation cooling are estimated to cost about $420,000 for engineering and installation. Although it is likely that a modification of this magnitude could be deferred to a refueling outage, immediate installation would require a forced outage. One licensee has estimated a potential loss of revenue of $24 million, based on a 2-month forced outage.

(2) The cost for installing full area automatic suppression and detection, sealing the open penetrations, and installing 1-hour-rated fire barrier and wraps in one fire area is estimated to be $3,350,000. Similarly, on the basis of a 2-month outage, lost revenue of $24 million is also a possibility.

D.2.2 PWR Case Study

The importance of this case study in terms of potential cost savings, therefore, is expected to be significant if such issues arise as a result of plant audits or inspections or self-examinations.

The objective of this case study is to demonstrate the feasibility of the approaches discussed in the earlier chapters of this report. The case study

described here is representative of a typical domestic PWR and does not represent a specific plant. The case study is designed to be as realistic as possible and at the same time allow a demonstration of various technical features in the framework.

A representative PWR emergency switchgear room (ESGR) is selected for this case study. The room is 15.2 m (50 ft) × 9.1 m (30 ft) and 4.6 m (15 ft) high. The room contains the power and instrumentation cables for the pumps and valves associated with motor-driven auxiliary feedwater (AFW) trains, all three high-pressure injection (HPI) trains, and both low-pressure injection (LPI) trains. The steam generator power-operated relief valves and the turbine-driven AFW trains are unaffected by a fire in this area. The power and instrument cables associated with safe-shutdown equipment are arranged in separate divisions and are separated horizontally by a distance, D. The value of D is varied for this case study.

A simplified elevation of the ESGR, illustrating critical cable locations, is shown in Figure D.6. The postulated ignition source is either a self-ignited cable (as a result of a fault) or cable ignition (as a result of a transient fire). The cable tray referred to as "tray A," located on the right side of the room at an elevation of about 2.3 m (7.5 ft) above the floor, is considered to be the source. Cables for the redundant train are contained in another tray (referred to as "tray B," the target). Tray B is separated from tray A by a horizontal distance, D, as shown in Figure D.6. The horizontal distance is varied in the sensitivity analysis. Three elevations are assumed for tray B. First, tray B is located 2.0 m (6.5 ft) above tray A (i.e., 0.3 m (1 ft) below the ceiling). This choice is made because, according to the FIVE methodology, tray B is in the ceiling jet sublayer when the ratio of height of target above fire source to the height from the fire source to ceiling is greater than 0.85 (6.5/7.5 = 0.87 in this case). Second, tray B is located 1.1 m (3.5 ft) above tray A. This implies that tray B is outside the ceiling jet sublayer but within the hot gas layer. Third, tray B is at the same elevation as tray A.

The configuration and scenario discussed here

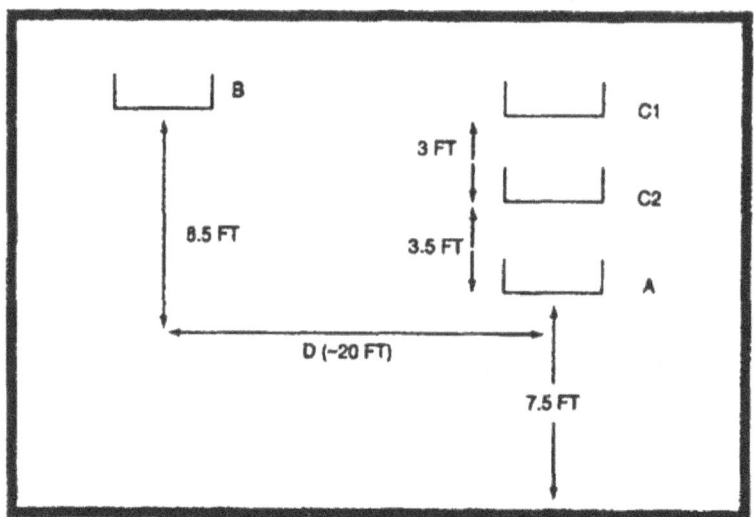

Figure D.6
Illustration of Critical Cable Locations in the Representative Emergency Switchgear Room
(Configuration 1)

will be analyzed using the FIVE, COMPBRN IIIe, and CFAST codes. In performing the analysis it is assumed that

- Other cable trays containing critical and non-critical cables are located directly above tray A.

- No combustible material intervenes between trays A and B.

- The ESGR has a small wall opening about 2.0 m (6.5 ft) high and about 0.2 m (0.7 ft) wide.

During a fire, most rooms will be isolated by the automatic closing of fire dampers and the shutdown of the ventilation system upon the detection of ignition. The assumption of an opening for the ESGR is a consideration that facilitates the use of both the COMPBRN and CFAST codes. An opening is needed to ensure no pressure buildup in the room.

The major fuel source is assumed to be insulation on cables installed in trays located in the upper section of the room. A typical PWR ESGR could contain about 13,608 kg (30,000 lb) of combustible insulation. Assuming that the cable

trays are 61 cm (24 in.) wide and 7.6 cm (3 in.) deep, the average insulation weight is about 44.6 kg/m (30 lb/linear foot). Hence, the assumed mass implies that there are about 305 m (1,000 ft) of cable trays in the room. Assuming that the heat of combustion of the insulation is about 20.6 megajoules (MJ)/kg (8850 Btu/lb), the total heat released is 280,050 MJ (265,500,000 Btu). For a floor area of 139.4 m² (1,500 ft²), the fire load is 2,010 MJ/m² (177,000 Btu/ft²). If the linear value of 15.1 MJ/m²/min (1,333 Btu/ft²/min) (908 MJ/m² (80,000 Btu/ft²) for the 1-hour American Society for Testing and Materials furnace test) developed by the National Fire Protection Association (NFPA) is used, the equivalent fire severity is about 133 minutes. This is considered to be a high fire severity.

All ESGRs contain fire protection systems. In this study, it is assumed that the ESGR contains smoke detectors and a manually actuated Halon system. The smoke detectors are spaced 9.1 m (30 ft) apart as recommended by NFPA 72E. The Halon system is capable of totally flooding the space with a 7-percent concentration of Halon and can maintain a concentration of at least 5 percent for a 10-minute period. Passive fire-retardant coatings on cable trays and conduits is not assumed for this study. It should be pointed out that the impact of actuation of the fire suppression system in controlling the fire cannot be treated by the conventional deterministic tools (i.e., FIVE, COMPBRN, and CFAST).

For the fire scenario discussed here, the only available equipment is associated with early decay heat removal and no injection system is available. The core, therefore, will eventually uncover as a result of cooldown and primary shrinkage unless one train of HPI is recovered.

The previous discussion conservatively assumes that all equipment within the fire area is damaged as a result of the fire. The probability of such an event occurring is evaluated through a detailed performance-based approach. This evaluation involves

- determination of fire initiating frequency

- determination of fire suppression probability both for automatic and manual suppression

- detailed modeling of fire propagation and the associated timing

This evaluation is performed using conventional PRA techniques: COMPBRN IIIe (EPRI NP-7282), FIVE methodology (EPRI TR-100370), and finally the CFAST code (Peacock et al., 1993b) developed by the National Institute of Standards and Technology. Various configurations for cable layout and combustible loading to obtain sufficient generic insights based on state-of-the-art analyses have been considered.

Fire Initiating Frequency

The representative fire area for this case study is similar to an ESGR. The room contains mainly cable trays, motor control centers (MCCs), and relay cabinets. The fire initiating frequency will account for self-ignited cables, externally ignited cables as a result of maintenance, welding activities, transient fires, and cabinet/MCC fires.

The frequency of self-ignited cable fires for IEEE fire-retardant cables has been a source of uncertainty in past PRAs. The Limerick PRA (NUS Corporation, 1983) reduces the self-ignited

cable fire frequencies by a factor of 1/3 to account for IEEE-rated cables. On the other hand, the FIVE methodology does not consider self-ignited cable fires (assigns a zero frequency). More recent PRAs assume that a short in a power cable (if not isolated) can produce enough heat to cause a sustained ignition. They report, on the basis of recent plant data, a frequency of 3.4E-8 per foot of cable tray per reactor-year (NUREG/CR-6144, Vol. 3, Pt. 1). Similarly, for externally ignited cables, a value of 2.9E-8 per foot of cable tray per reactor-year is reported. The associated error factor for these estimates described by a lognormal distribution is about 3.0. The frequency of a large fire initiated in an MCC is estimated to be 1.8E-5 per reactor-year with an error factor of 10. A higher frequency is reported for relay cabinets, namely, 6.2E-5 per reactor-year with an error factor of 3. Finally, the probability of transient fires for areas similar to the ESGR is reported to be about 1.4E-3 per reactor-year with an error factor of about 3.

This case study assumes an ESGR with 1,838 ft of cable trays, five MCCs, and five relay cabinets, and considers transient fires to obtain a mean fire initiator frequency of about 2.0E-3 per reactor-year with an error factor of about 4. This frequency does not include the area ratio fraction for transient fuels. The most credible transient fires must be within 3 ft of the source cables in order to ignite the cables (NUREG/CR-4832, vol.9, P3-76 to 3-80). This assumption results in a critical area ratio fraction for transient fires of 0.2; that is, the transient fire initiator frequency is to be reduced by a factor of 5. This will result in a fire initiator frequency of 8.8E-4 per reactor-year with an error factor of about 3.

Determination of Fire Suppression Probability

As described earlier, this area is equipped with a manually actuated Halon system and a smoke detector system. A fire in this area is most likely to be detected either by smoke detectors or by an employee. The detection time for similar areas (NUREG/CR-4230) is expected to be less than 2 minutes. Once the fire is detected, it may be controlled by manual actuation of the Halon system. The time required for this manual action is estimated to be less than 15 minutes, and the unavailability of the Halon system, from the same reference, is estimated to be 0.08 with an error factor of about 2. Finally, if the fire is not controlled by the Halon system, it will eventually be extinguished by the fire brigade. The empirical data for the probability of failure of the fire brigade to suppress the fire, P_{ns}, in time, t, is expressed through a Weibull probability distribution (NUREG/CR-6144, Vol. 3, Pt. 1); that is,

$$P_{ns} = \exp\left[-\left(t/\xi_c\right)^{\sigma_c}\right] \qquad (D-1)$$

where, typically, values of ξ_c and σ_c are 20 minutes and 0.5 (unitless), respectively, for most areas in nuclear power plants.

Reliability models can be used to arrive at the overall failure probability of suppression accounting for detection, autosuppression, and manual suppression as a function of time. Figure D.7 shows the failure probability of suppression for this case study with the associated uncertainty limits. Note that the break point in the curves is a result of the finite timing for manual actuation of Halon systems. Figure D.8 shows a similar graph wherein one considers an automatic suppression system with fast actuation of less than 2 minutes instead of the manually actuated Halon system.

From the information presented in these figures and considering the fire initiator frequency of about 8.8E-4 (as discussed previously), a total damage probability of 1.2E-5 (corresponding to failure of suppression of 1.5E-2 in 1 hour) can be obtained. It therefore has to be shown from the fire propagation modeling that the redundant equipment is not damaged within 1 hour to ensure a CDF below 1.2E-5 per reactor-year for this fire scenario. Note that the 1-hour time limit is imposed mainly by the non-suppression probability curve given in Equation D-1 as reported in NUREG/CR-6144. Use of other non-suppression probability curves may result in much shorter time limits than 1 hour.

Fire Propagation Models

The fire propagation models discussed next are used to determine the maximum cluster of cables and the resulting peak burning rate that most likely would not damage the redundant cables in less than 1 hour. Three different methodologies were used—the FIVE method, COMPBRN IIIe, and the modified CFAST code—to provide a spectrum of different results. The utilization and the results of each model are as follows.

FIVE Analysis

The FIVE screening methodology was applied to determine the magnitude of the effective fire intensity that can damage redundant cables that are separated by a distance (e.g., about 6.1 m (20 ft)). A heat loss factor of 0.7 is included in the FIVE method. The FIVE fire screening methodology considers three general scenarios. For the present case in which the target cable is separated from the source cable by a horizontal distance of about 6.1 m (20 ft), analyses were performed using Worksheet 2 (see Table D.1) outlined in EPRI TR-100370. Three cases were considered. In Case 1, the target is located in the ceiling jet sublayer region, that is, 0.3 m (1 ft) below the ceiling. For this situation, an effective peak fire intensity must be estimated from which the ceiling jet temperature is evaluated. If the ceiling jet temperature exceeds the threshold damage temperature of the target (assumed to be 643 K (698 °F) in this study), the scenario being evaluated does not pass the basic FIVE screening process. The results of the FIVE analyses are shown in Table D.2. At a separation distance of about 6.1 m (20 ft), the critical fire intensity is about 6.5 MW (22.2 million Btu/hr), above which the ceiling jet temperature exceeds the assumed cable damage temperature. When the separation distance is reduced to about 3 m (10 ft), the critical fire intensity is reduced to about 3.5 MW (11.9 million Btu/hr).

In Case 2, the target elevation is reduced to 1.2 m (4 ft) below the ceiling. The target is outside the ceiling jet sublayer, but within the hot gas layer. For this situation, FIVE estimates the total energy release needed to raise the average hot gas layer temperature to the threshold damage temperature. This quantity then is compared with the total energy available in the exposure fire fuel. If the total energy available exceeds the energy needed

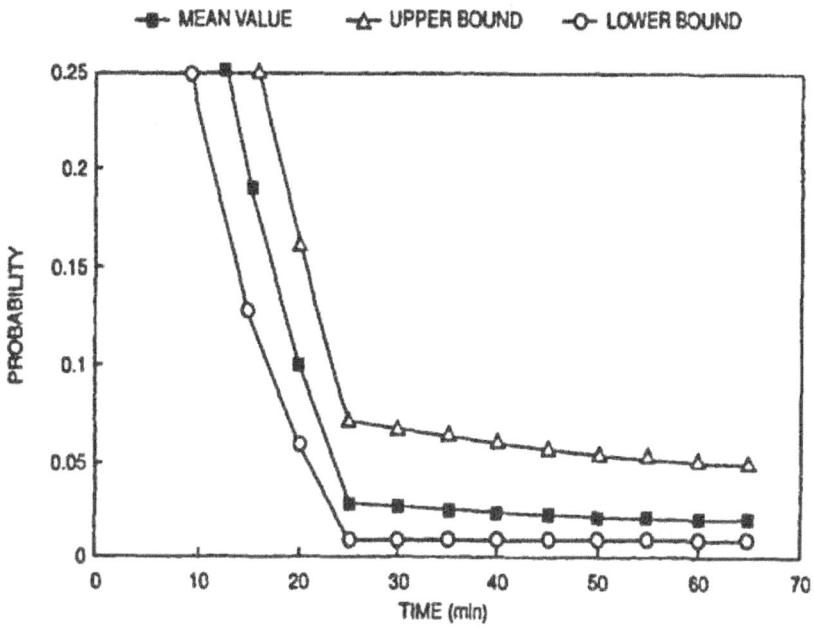

Figure D.7
Failure Probability of Suppression vs. Time for Manually Actuated Halon Systems

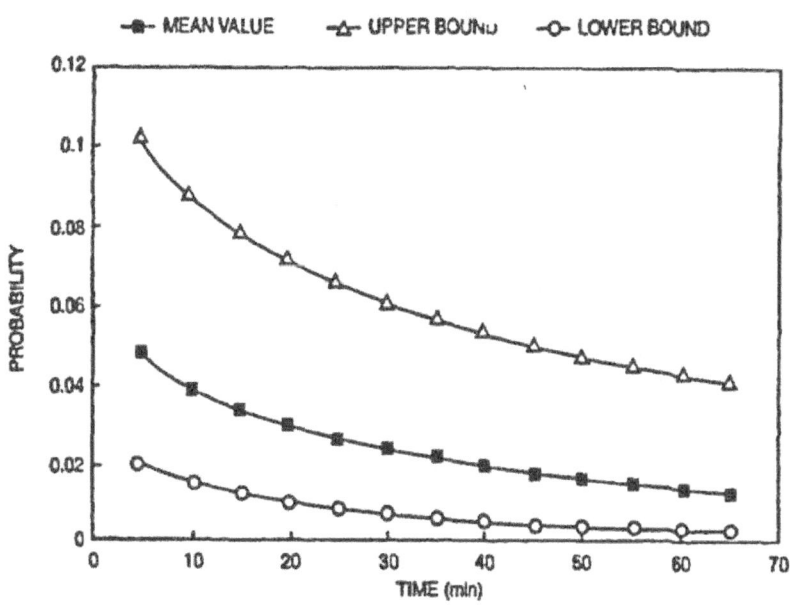

Figure D.8
Failure Probability of Suppression vs. Time for Automatically Actuated Suppression

to raise the hot gas layer temperature to the damage temperature, the scenario does not pass the screening process. For the present case, the total energy needed to raise the hot gas layer temperature from an initial 300 K (80.6 °F) to the damage threshold of 643 K (698 °F) is about 286 MJ (0.27 million Btu). The total energy available, however, depends on the time of exposure to fire. The total energy available is about 3,150, 5,850, and 6,300 MJ (3.3, 5.5, and 6.0 million Btu) during a 15-minute period at fire intensities of 3.5, 6.5, and 7 MW (11.9, 22.2, and 23.9 million Btu/hr), respectively. The total energy available is based on an adiabatic heating of the gas (i.e., no heat loss from the fire source). Since the total energy available is much larger than the total energy needed, none of the cases can pass the screening process. This calculation is not affected by the separation distance.

In Case 3, the target elevation is further reduced to the elevation of the source, that is, 2.29 m (7.5 ft) below the ceiling. According to the FIVE methodology, the target is still within the hot gas layer region. The result is identical to that of Case 2.

An example of FIVE-predicted results, for 6.5 MW (22.2 million Btu/hr), is presented in the form of Worksheet 2 as Table D.1.

In summary, the FIVE methodology predicts that an effective fire source intensity of about 6.5 MW (22.2 million Btu/hr) is required to damage cables that are separated by 6.1 m (20 ft) if the cables are in the ceiling jet layer. Similarly, a source fire of 3.5 MW (11.9 million Btu/hr) is sufficient to damage redundant cables that are 3 m (10 ft) apart. When the cables are in the hot gas layer, FIVE does not differentiate between the various separation distances and predicts a total of 286 MJ (0.27 million Btu) heat release for cable damage, therefore requiring more detailed calculation to be performed by COMPBRN.

COMPBRN Analyses

Point estimates of cable damage and ignition times were determined by using the COMPBRN IIIe computer code. The code requires a number of model parameters and cable physical properties as input data. In the present study, all model parameters were the default values recommended by the code. Most of the cable property parameters are the same as those used in EPRI NP-7282. Only three property values were modified as shown in Table D.3 (NUREG/CR-4230; NUREG/CR-4679).

The cable damage and ignition temperatures are assumed to be 643 and 733 K (698 and 860 °F), respectively. Five cases, including a base case (Case 1) and four sensitivity studies, were examined. The cases are summarized in Table D.4. In the base case, it is assumed that the source tray A (referred to as "pilot fire" in COMPBRN analysis) is located at an elevation of 2.29 m (7.5 ft) above the floor. Two cable trays are located directly above tray A. Tray C1 is 1.98 m (6.5 ft) and C2 is 1.07 m (3.0 ft) above the pilot fire. Since the flame height predicted by the COMPBRN code is about 1 m (3.3 ft), tray C2 is likely to be within the pilot flame region and tray C1 to be within the flame region of tray C2 if tray C2 is burning. Length of the pilot fire is assumed to be 1.2 m (4 ft) in the base case (i.e., two elements according to the nodalization modeled in this study). Each element is 0.6 m (2 ft) long and 0.6 m (2 ft) wide. The target tray (tray B) is separated from tray A by a horizontal distance of 6.1 m (20 ft) and is located at an elevation of 4.27 m (14 ft) above the floor (about 1.98 m (6.5 ft) higher than tray A and 0.3 m (1 ft) below the ceiling). An opening about 2 m (6.5 ft) high and about 0.2 m (0.7 ft) wide is assumed for this case.

COMPBRN predicted that trays C2 and C1 (located directly above tray A) are damaged at 2 and 4 minutes, respectively. At the time of damage, one element of each of the two trays is also ignited. The fire spreads longitudinally along the trays and, in about 8 minutes, six elements of each tray are ignited. The large fire causes damage and ignition of tray B, which is separated by a horizontal distance of 6.1 m (20 ft). COMPBRN predicts transient burning rate, total heat release rate, and the temperatures of tray B and the hot gas at the time when tray B is damaged. The burning rate is about 0.25 kg (0.55 lb) per second and the corresponding heat release rate is about 4.8 MW (16.4 million Btu per hour).

Table D.1 FIVE Worksheet 2 (Fire Intensity = 6.5 MW)

PWR ESGR 20-ft separation, heat release = 6.5 NW, case-1

2   Worksheet 2: Target - outside-plume Scenario

1	target damage threshold temperature	643.0	K
2	height of target above fire source	2.0	M
3	height from fire source to ceiling	2.3	M
4	ratio of target height/ceiling ht.	.9	
5	long. distance from fire to target	6.1	M
6	long. distance to height ratio	2.7	M
7	enclosure width	9.1	M
8	height to width ratio	.2	
9	peak fire intensity	6500.0	KW
10	fire location factor	1.0	
11	effective heat transfer rate	6500.0	KW
12	plume temperature rise at ceiling	2195.5	C
13	ceiling jet temp. rise factor at tg	.2	
14	ceiling jet temp. rise at target	342.5	C
15	critical temperature rise at target	343.0	C
16	critical-ceiling jet t rise at targ	.5	C

Critical temp rise > ceiling jet temp rise !
Box 16 becomes the critical average temperature rise. The following calculations are used to evaluate the critical combustible loading needed, to achieve this average temperature rise.

17	Qnet/V to achieve temp rise in 16	.6	KJ/M3
18	Calculated enclosure volume, V	318.6	M3
19	calculated critical Qnet	186.0	KJ
20	estimated heat loss fraction	.7	
21	estimate of critical Qtot	619.9	KJ
22	estimate of actual Qtot	5850000.0	KJ

This scenario does not pass the screening procedure!
Farther analysis is required

PWR ESGR 20-ft separation, heat release = 6.5 MW, case-2

2   Worksheet 2: Target - outside-plume Scenario

1	target damage threshold temperature	643.0	K
2	height of target above fire source	1.1	M
3	height from fire source to ceiling	2.3	M
4	ratio of target height/ceiling ht.	.5	

Target is beneath the ceiling jet sublayer
Go to Box 14

Table D.1 (cont'd.)

5	long. distance from fire to target	6.1	M
6	long. distance to height ratio	2.7	M
7	enclosure width	9.1	M
8	height to width ratio	.2	
9	peak fire intensity	6500.0	KW
10	fire location factor	1.0	
11	effective heat transfer rate	6500.0	KW
12	plume temperature rise at coiling	2195.5	C
13	ceiling jet temp. rise factor at tg	.2	
14	ceiling jet temp. rise at target	.0	C
15	critical temperature rise at target	343.0	C
16	critical-ceiling jet t rise at targ	343.0	C

Critical temp rise > ceiling jet temp rise !
Box 16 becomes the critical average temperature rise. The following calculations are used to evaluate the critical combustible loading needed to achieve this average temperature rise.

17	Qnet/V to achieve temp. rise in 16	269.1	KJ/M3
18	Calculated enclosure volume, V	318.6	M3
19	calculated critical Qnet	85727.3	KJ
20	estimated heat lose fraction	.7	
21	estimate of critical Qtot	285757.8	KJ
22	estimate of actual Qtot	5850000.0	KJ

This scenario does not pass the screening procedure!
Farther analysis is required!

PWR ESGR 20-ft separation, heat release = 6.5 KW, case-3

2   Worksheet 2: Target - outside-plume Scenario

1	target damage threshold temperature	643.0	K
2	height of target above fire source	.0	M
3	height from fire source to ceiling	2.3	M
4	ratio of target height/ceiling ht.	.0	

Target is beneath the ceiling jet sublayer
Go to Box 14

5	long. distance from fire to target	6.1	M
6	long. distance to height ratio	2.7	M
7	enclosure width	9.1	M
8	height to width ratio	.2	
9	peak fire intensity	6500.0	KW
10	fire location factor	1.0	
11	effective heat transfer rate	6500.0	KW
12	plume temperature rise at ceiling	2195.5	C
13	ceiling jet temp. rise factor at tg	.2	
14	ceiling jet temp. rise at target	.0	C
15	critical temperature rise at target	343.0	C
16	critical-ceiling jet t rise at targ	343.0	C

Table D.1 (cont'd.)

Critical, temp > ceiling jet temp rise !
Box 16 becomes the critical average temperature rise. The following calculations are used to evaluate the critical combustible loading needed to achieve this average temperature rise.

17	Qnet/V to achieve temp rise in 16	269.1	KJ/M3
18	Calculated enclosure volume, V	318.6	M3
19	calculated critical Qnet	85727.3	KJ
20	estimated heat loss fraction	.7	
21	estimate of critical Qtot	288757.8	KJ
22	estimate of actual Qtot	5850000.0	KJ

This scenario does not pass the screening procedure!
Farther analysis is required

PWR ESGR 10-ft separation, heat release = 6.5 NW, case-1a

2   Worksheet 2: Target - outside-plume Scenario

1	target damage threshold temperature	643.0	K
2	height of target above fire source	2.0	M
3	height from fire source to ceiling	2.3	M
4	ratio of target height/ceiling ht.	.9	
5	long. distance from fire to target	3.0	M
6	long. distance to height ratio	1.3	M
7	enclosure width	9.1	M
8	height to width ratio	.2	
9	peak fire intensity	6500.0	KW
10	fire location factor	1.0	
11	effective heat transfer rate	6500.0	KW
12	plume temperature rise at ceiling	2195.5	C
13	ceiling jet temp. rise factor at tg	.2	
14	ceiling jet temp. rise at target	543.7	C
15	critical temperature rise at target	343.0	C
16	critical-ceiling jet t rise at targ	-200.7	C

Ceiling jet temperature rise exceeds the damage threshold temperature !
This scenario does not pass the screening procedure !

Table D.2 Summary Results From FIVE Analyses

Effective Fire Intensity kW	Ceiling Jet Temperature K	Target Damage Temperature K	Separation Distance ft
3500	526	643	20
6500	643	643	20
7000	660	643	20
3500	660	643	10
6500	843	643	10
7000	871	643	10

Table D.3 Modified Parameters Used for COMPBRN IIIe

Heat value	26.5 MJ/kg
Surface control burning rate constant	0.4E-6 kg/J
Fraction of flame heat released as radiation	0.48

In Case 2, the size of the pilot fire is reduced to 0.6 m × 0.6 m (2 ft × 2 ft). However, the damage and ignition of the target are only delayed by 1 minute. Apparently the size does not have a significant effect on fire growth. The total heat-release rate at the time of damage is about 4 MW (~13.6 million Btu per hour).

Case 3 assumes that the ESGR has no openings. COMPBRN modeled this scenario as a closed-door fire; that is, the entire room is in the hot gas layer. In this situation, the target damage time is delayed to 12 minutes, at which time the total heat release rate is about 8.2 MW. However, no ignition of the target is predicted because the target temperature does not reach the assumed cable ignition temperature (733 K (860 °F)). This is probably due to the modeling of a closed-door fire in the COMPBRN code.

The elevation of the target is reduced to 2.29 m (7.5 ft) in Case 4. This is the same as the elevation

of the pilot fire. The change of elevation has no effect on the target. The target damage and ignition times are identical to that of the base case.

Finally, tray C1 is removed from the analysis in Case 5. Only tray C2 is located within the pilot flame region and is damaged and ignited at 2 minutes, similarly to the base case. Because no other cable tray is located above tray C2, COMPBRN predicts no upward fire propagation. The fire in tray C2 propagates slowly along the tray. At 10 minutes, three elements of tray C2 have ignited and the total heat-release rate is 1.8 MW (6.1 million Btu per hour). At 14 minutes, only one element of tray C2 is still burning and the total heat-release rate is reduced to 1.1 MW (3.7 million Btu per hour). COMPBRN predicts no damage to tray B (target) because of the low heat-release rate. The results of the base case and sensitivity studies are summarized and compared in Table D.4.

Table D.4 Summary of COMPBRN Results

Tray	Case 1		Case 2		Case 3		Case 4		Case 5	
	D	I	D	I	D	I	D	I	D	I
I. Damaged (D) and Ignition (I) Time (minutes)										
A (Source)	0	0	0	0	0	0	0	0	0	0
C2	2	2	2	3	2	2	2	2	2	2
C1	4	4	5	5	4	4	4	4	-	-
B (Target)	8	9	9	10	12	No	8	9	No	No
II. Total Heat Release Rate at the Time of Target Damage										
Q, MW	4.8		4.0		8.2		4.7		1.8*	
III. Description of Cases										
Pilot fire size (ft × ft)	4 × 2		2 × 2		4 × 2		4 × 2		4 × 2	
Door	Open		Open		Closed		Open		Open	
Trays above pilot fire	C1 and C2		C1 and C2		C1 and C2		C1 and C2		C2 only	
Target elevation (m)	4.27		4.27		4.27		2.29		4.27	

*Maximum heat release rate with no damage to target cables.

COMPBRN analyses predict that the effective fire intensity, capable of damaging redundant cables separated by 6.1 m (20 ft), is about 4 MW (~13.6 million Btu per hour) for the representative configuration. COMPBRN also predicts damage time of about 12 minutes. These results are obtained when a sufficiently large opening is assumed and therefore oxygen is always available for combustion in the room. Furthermore, COMPBRN results show that the cluster of two cable trays in one side of the room will result in a peak burning rate of about 1.8 MW (6.1 million Btu per hour), which is not sufficient to damage cable trays separated by 6.1 m (20 ft).

CFAST Analyses

A modified version of the CFAST code, which accounts for radiation heat transfer to a target, was utilized for this case study. CFAST requires the heat-release rate of the source fire as input. To arrive at a meaningful heat-release rate for the fire source, a radiation model was implemented in the MATHEMATICA computer package. The heat flux at distance, D, due to radiation, was modeled using the following equation:

$$q'' = (\text{Cos } \theta) (q_f) [F/(4 \pi D^2)] \text{ W/m}^2$$

where

θ = the angle between the tray and the fire source

q_f = the heat-release rate of the fire source

F = the fraction of the heat-release rate radiated (set to 0.48)[*]

D = the separation distance of the target

To damage the IEEE-rated cables, an external heat flux of about 10 kW/m² at the target cables was assumed. The 10 kW/m² external heat flux is reported in several studies (NUREG/CR-4679; U.S. Department of Transportation, 1983) as a sufficient heat flux to damage cables. For various separation distances, D, the corresponding value of q_f was estimated. The values of q_f of interest ranged from 2 to 5 MW for damaging redundant cable trays at various distances, D. On the basis of this insight, the CFAST computer code was utilized with the source fire of 1 MW (3.4 million Btu per hour), 2 MW (6.8 million Btu per hour), and 3 MW (10.2 million Btu per hour) to assess the damage time for target cables. Extrapolation of the results allows sensitivity of target damage time as a function of the heat-release rate of the source fire.

The CFAST code was then utilized to model the specific geometry of the case study, with the heat-release rate for the fire source of 1 MW, 2 MW, and 3 MW. The peak heat-release rate of the fire source (i.e., 1 MW, 2 MW, and 3 MW) was reached through a linear growth taking 1, 2, and 3 minutes, respectively. The hot layer temperature, the radiative and convective heat transfer calculated by CFAST, was used in a transient conduction model for a thin slab to estimate the target surface temperatures. Figures D.9, D.10, and D.11 show the cable surface temperature for a 1-, 2-, and 3-MW fire as a function of time.

These figures are for a separation distance of 6.1 m (20 ft) and for target cable trays located inside the hot layer. CFAST models the ceiling jet layer; however, none of the targets appear to be in the ceiling jet.

Considering the critical damage temperature of 643 K (698 °F) and the extrapolation of the results shown in these figures, a fire of more than

---

*To be consistent with the COMPBRN runs.

3 MW (10.2 million Btu per hour) is required to damage the target cables at a 6.1-m (20-ft) separation in about 1 hour. Since the hot layer temperature and, therefore, convective heat transfer, do not vary with separation distance, the only consideration is the radiative heat transfer, which is proportional to $1/D^2$. For separation distances greater than 3 m (10 ft) and less than 6.1 m (20 ft), the hot layer temperature is a better indication of damageability for the cables. This, in turn, limits the maximum size of the source fire to 3 MW (10.2 million Btu per hour) to avoid damage to the target cables.

Results from FIVE, COMPBRN and CFAST are compared in Appendix C.

D.2.3 Summary

The case study selected deals with a fire area similar to the emergency switchgear room at a PWR plant where the 6.1-m (20-ft) separation criterion is not met; that is, the actual separation between the cables associated with redundant trains is 4.6 m (15 ft). A large fire, damaging all the equipment in this area, will eventually lead to core damage if repair is not credited. The performance-based approach demonstrates the use of the available fire methodologies. Application of the three different methodologies—FIVE, COMPBRN, and CFAST—resulted in limits on peak heat-release rates varying from 6.5 MW (22.2 million Btu per hour) down to 3 MW (10.2 million Btu per hour) to cause damage to redundant cable trays. The damage time also varied from 10 minutes up to 1 hour. A fire of 3-MW magnitude was estimated to take about 1 hour to damage redundant cables that are separated by more than 3 m (10 ft). It was also shown that a fire of 2 MW (6.8 million Btu per hour) or less of the heat-release rate will not damage the redundant cable trays. Considering a heat of combustion of 25 MJ/kg (~107,000 Btu/lb) and a surface-controlled specific mass loss rate of about 3 g/m²-sec (2.21 lb/ft²-hr) for cables that pass the Institute of Electrical and Electronics Engineers (IEEE) test-rated cables, a 15-m (50-ft) cable tray, 0.6 m (2 ft) wide will have an effective heat release of about 0.9 MW (3 million Btu per hour). (ses Section C.1 for further justification of this assumption.) Therefore, the source fire

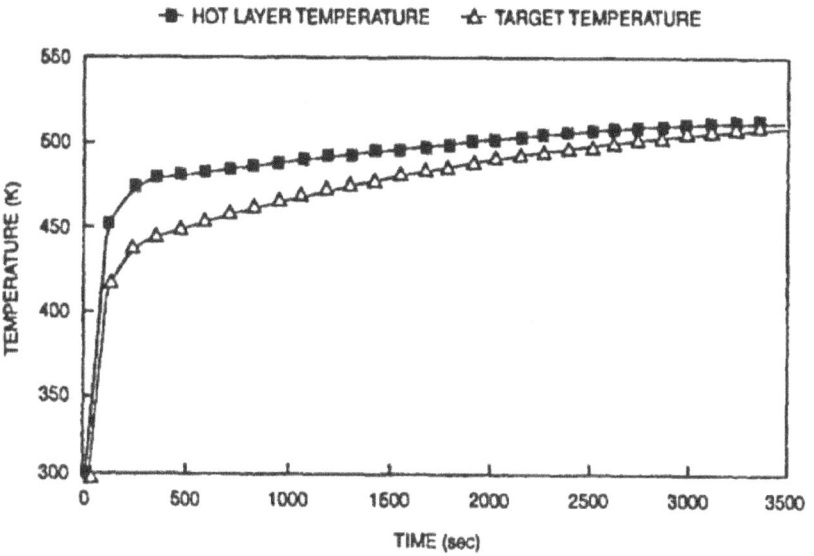

Figure D.9
1-MW Fire Source Target and Hot Layer Temperature

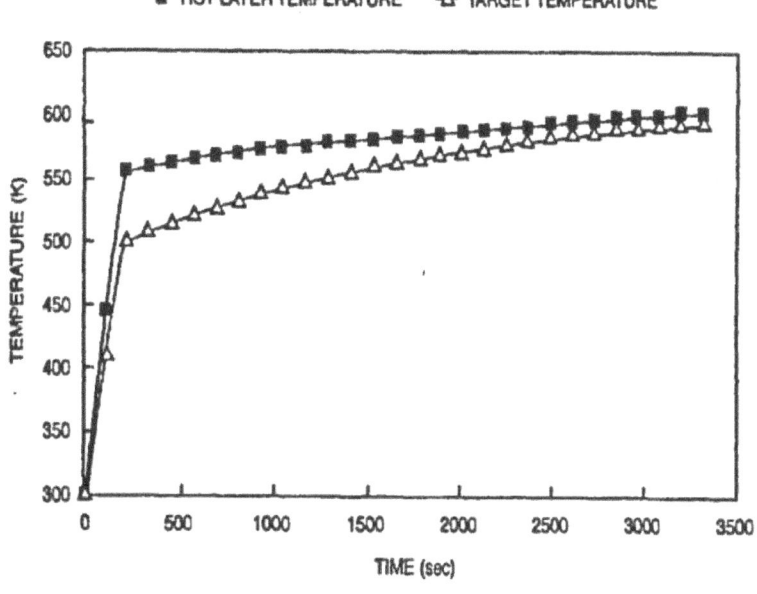

Figure D.10
2-MW Fire Source Target and Hot Layer Temperature

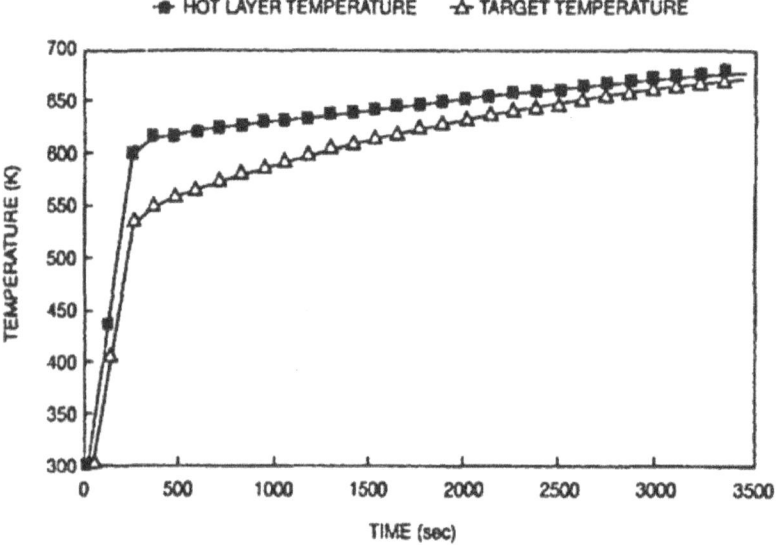

Figure D.11
3-MW Fire Source Target and Hot Layer Temperature

limited to a maximum cluster of three cable trays is expected to produce a heat-release rate of less than 2 MW (6.8 million Btu per hour).

The dominant factor for all these methodologies for predicting damage to cables that are separated by 6.1 m (20 ft) is the effective intensity of the fire source, not the total combustible loading in the fire area. All fire sources with the effective intensity less than the critical fire severity* were screened out because of the low probability of suppression failure. The critical fire severity is determined by use of the available fire propagation methodologies. The following insights can be drawn from this case study:

(1) FIVE can determine the peak heat-release rate of a fire to cause damage at a target in the ceiling jet layer at various separation distances.

(2) FIVE can screen out those areas with a low combustible loading for targets within the hot layer. FIVE assumes that the hot layer thickness is the distance between the lowest exposure fire and the ceiling. Therefore, it is too conservative for fires near the ceiling and not conservative enough for fires near the floor.

(3) COMPBRN IIIe is capable of simulating small- to moderate-sized fires. For large fires (greater than 4 MW (13.6 million Btu per hour) in this case study) and for fast-growing fires, the results of COMPBRN are not consistent with those from CFAST.

(4) CFAST is capable of simulating larger fires; however, the fire heat-release rate is to be estimated by the user from either experimental data or actual fire events.

The best estimate of the critical fire severity calculated for this case study is a fire source with heat output of 3 MW (10.2 million Btu per hour). The performance-based analysis shows that if the maximum cluster of source cables results in a

---

* The critical fire severity is defined as the effective intensity that is predicted to cause damage to separated, redundant cables at 1 hour after fire initiation. This 1-hour duration is case specific and includes consideration of the reliability and effectiveness of the suppression mechanisms (both manual and automatic), as well as the conditional core-damage probability.

heat-release rate of 2 MW (6.8 million Btu per hour) or less, then the redundant cables will not be damaged even if they are separated by less than 6.1 m (20 ft) (e.g., 4.6 m (15 ft)). However, if the heat-release rate is about 3 MW (10.2 million Btu per hour) or more, the CDF caused by fire is estimated to be greater than 1E-5. For this case study, the quantitative risk-informed approach estimates a ΔCDF of 5E-6 between the assumed configuration (4.6-m (15-ft) separation) and a configuration in compliance with Appendix R, Section III.G (protection of safe-shutdown capability). The fire propagation and results depend greatly on the specific configuration of the case being analyzed. The reader is reminded that the importance of this case study relies on the approach and demonstration of the methodology, not on the final case-specific conclusions.

D.3 ANALYSIS OF THE 72-HOUR CRITERION TO REACH COLD SHUTDOWN

This case study examines the Appendix R requirement to achieve and maintain cold shutdown within 72 hours of a fire. It is generally based on fire area AC as modeled in the LaSalle fire PRA. The feasibility of an alternative approach to prescriptive compliance is explored using two levels of modeling resolution. Case 1 adopts the conservative modeling used by the LaSalle PRA. No credit is taken for any operator recovery actions. Case 2 models key operator recovery actions to reestablish and maintain the main condenser heat sink to allow extensive repairs to the residual heat removal (RHR) system. The CDF associated with the alternative approach is compared with the CDF assuming prescriptive compliance for each case. An uncertainty analysis is also performed to examine the distribution of the ΔCDFs. This CDF difference can be used as one input in the assessment of an alternative approach to a prescriptive requirement.

The requirement to achieve and maintain cold shutdown within 72 hours of a fire is stated in two sections of Appendix R. Section III.G, "Fire Protection of Safe Shutdown Capability," subsection 1.b states:

1. Fire protection features shall be provided for structures, systems, and components important to safe shutdown. These features shall be capable of limiting fire damage so that:

(b) Systems necessary to achieve and maintain cold shutdown from either the control room or emergency control station(s) can be repaired within 72 hours.

Section III.L, "Alternative and Dedicated Shutdown Capability," subsections 1.d and e state:

1. Alternative or dedicated shutdown capability provided for a specific fire area shall be able to:

(d) achieve cold shutdown conditions within 72 hours; and

(e) maintain cold shutdown conditions thereafter.

Furthermore, Section III.L.5 states:

5. Equipment and systems comprising the means to achieve and maintain cold shutdown conditions shall not be damaged by fire; or the fire damage to such equipment and systems shall be limited so that the systems can be made operable and cold shutdown can be achieved within 72 hours. If such equipment and systems used prior to 72 hours after the fire will not be capable of being powered by both onsite and offsite electric power systems because of fire damage, independent onsite power system shall be provided. Equipment and systems used after 72 hours may be powered by offsite power only.

The purpose of this requirement is to limit the extent of fire damage to the systems that are necessary to achieve cold shutdown. The requirement in Section III.G has been clarified in later NRC documentation to require the capability to be in cold shutdown within 72 hours, as opposed to actually requiring cold shutdown within that time. However, the capability to reach stable shutdown* by alternative methods that require more than 72 hours or offsite power or both should be considered if it can be demonstrated that the analytical assumptions are appropriate and the additional risk is minimal.

D.3.1 Background

As discussed briefly in Chapter 3, several licensees have requested exemptions from the requirement to achieve cold shutdown within 72 hours. This review did not identify a request for exemption from the 72-hour cold shutdown criterion for any BWR. The available decay heat removal systems in a BWR, RHR shutdown cooling, the power conversion system, or alternate shutdown cooling (using the safety/relief valve system and low-pressure coolant injection), have the capability to bring the plant to cold shutdown well before 72 hours. The intent was to use a PWR to illustrate this case study. A search of the public document room (PDR) identified several PWR exemption requests. In addition, a previous Brookhaven National Laboratory fire study for shutdown and low power operations at Surry Unit 1 (NUREG/CR-6144, Vol. 3) identified fire areas with both trains of RHR affected. Finally, the PWR can stay in hot shutdown for long periods of time, especially without offsite power.

However, when the detailed PDR information was received, the reasons for the various exemption requests were clarified. For example, several Babcock & Wilcox (B&W) PWR licensees have requested and received exemptions from the requirement in Section III.L to achieve cold shutdown within 72 hours independent of offsite power. The B&W design is such that pressurizer spray capability depends on operation of the reactor coolant pumps, which, in turn, requires offsite power. If pressurizer spray is not available, depressurization of the reactor and subsequent cooldown are determined by the rate

*Stable shutdown can be less restrictive than the technical specification definition for cold shutdown.

of heat loss from the pressurizer to the containment environment. A further restriction on the cooldown rate is typically imposed to avoid formation of steam in the upper head. Using this cooldown restriction and a conservative analysis, the Rancho Seco licensee calculated that 205 hours would be required to achieve cold-shutdown conditions, assuming offsite power was unavailable.

The licensee for Beaver Valley Unit 1, a Westinghouse PWR, also received an exemption from Section III.L. The licensee proposed an alternative shutdown capability that was independent of the RHR system and offsite power. Cold-shutdown conditions can be achieved and maintained by going to a solid steam generator. In this method, the steam generator receives makeup water from the auxiliary feedwater system and drains to the main condenser via the steam bypass dump valves. The licensee has estimated that this process would require about 127 hours to achieve cold shutdown. The exemption was requested on the basis of a deterministic engineering analysis.

In the aforementioned examples, unusual system configurations and success paths were used. The available PWR PRAs (both internal events and fire) do not model these alternative paths to cold shutdown. The evaluation of an alternative approach to a prescriptive compliance measure requires detailed PRA modeling for each scenario. Although this exemption would have been a good case to illustrate the use of a risk-informed approach, these alternative cold shutdown paths could not be evaluated for the purposes of this study without extensive additional modeling.

D.3.2 BWR Case Study

As stated previously, there do not appear to be any requests for exemption from the 72-hour criterion for BWRs. This implies that for every fire area, BWRs have an undamaged train of cold-shutdown systems or that any fire damage can be repaired in a timely fashion. The LaSalle plant is a typical example. In the LaSalle PRA (NUREG/CR-4832, Vol. 9), no fire areas contain both RHR trains. Additional random failures of the undamaged RHR train are generally required to cause core damage. Therefore, the LaSalle plant conforms to the Appendix R requirement to have the capability to be in cold shutdown within 72 hours.

To illustrate an alternative approach, the LaSalle PRA analysis of fire area AC is used as a surrogate. Fire area AC is the cable shaft room adjacent to the Unit 2, Division 2, essential switchgear room. This room is located in the auxiliary building. This case study postulates that fire area AC contains equipment associated with both trains of RHR. The postulated damage in fire area AC is extensive, and it will take more than 72 hours to restore one RHR train. Prescriptive compliance assumes that one RHR train will be removed from area AC or protected. The alternate approach does not credit plant modifications. Two levels of modeling resolution are examined for this case study. In accord with the conservative modeling assumptions of the LaSalle PRA for a fire in area AC, Case 1 does not credit any operator recovery actions. Early reactor pressure vessel (RPV) injection is maintained for most of the sequences. However, random or fire-induced losses of decay heat removal are assumed to cause containment failure due to overpressurization in about 27 hours. The resulting harsh environment in the reactor building can fail RPV injection and cause late core damage.

Case 2 uses a finer level of modeling resolution. Manual recovery actions to reestablish containment heat removal are included in the event tree model.

The difference in core-damage frequency (ΔCDF) between the prescriptive compliance case and the alternative approach for fire area AC is examined for each level of modeling resolution.

Case 1—Conservative Modeling Assumptions

The prescriptive compliance case assumes that RHR train B is either removed from fire area AC or suitably protected. Protection could entail separation or fire barrier(s). Fire-related damage fails all or parts of the following LaSalle systems:

- main feedwater

- condenser (due to main steam isolation valve (MSIV) closure)

- train A of RHR including shutdown cooling, suppression pool cooling, containment spray, and low-pressure coolant injection

- containment venting

The sequences associated with fire area AC for both the prescriptive compliance and the alternative cases are presented in Figure D.12. This event tree uses the conservative modeling assumptions of the LaSalle PRA. The initiator, FIRE-AC, is the estimated frequency of a significant fire in room AC. The first branch of the tree examines the likelihood of early RPV injection (top event E-INJ). Sequence 10 represents early core damage caused by random failures of early RPV injection. Given successful early RPV injection, the containment heat removal function is examined (top event CHR). Sequence 1 is a successful end-state and represents one path to cold shutdown after a fire in room AC. If containment heat removal is not available, the core decay heat will cause a containment overpressure failure in about 27 hours. The tree estimates the probability of the failure for continued RPV injection given the severe environment in the reactor building caused by primary containment failure (top event L-INJ). Sequence 8 represents continued injection. Although the reactor is not in cold shutdown, it is considered to be a successful end-state. Sequence 9 models the loss of RPV injection after containment failure.

The quantified event tree for the conservative modeling case is presented as Figure D.13. The estimation of each top event is discussed below.

Initiator FIRE-AC

The probability of damage to critical equipment in a fire area can simply consist of an estimate of the initiator frequency of a significant fire in conjunction with the assumption that all components in the fire area are failed. If warranted, this simplification can be replaced by a more realistic analysis that can refine the fire initiator frequency or examine fire propagation

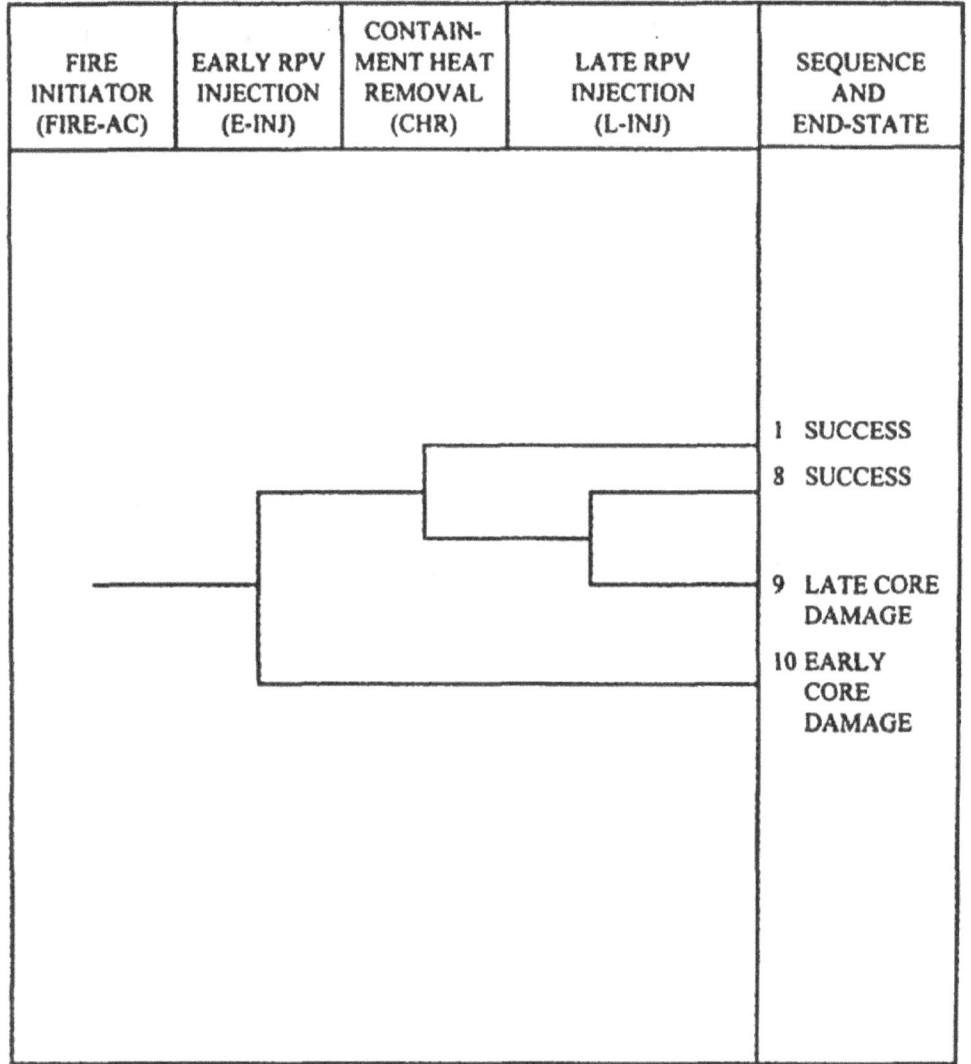

Figure D.12
72-Hour Case Study—Event Tree for Case 1 Event

and suppression probabilities (see Chapter 4).

The fire analysis is adapted from the LaSalle fire PRA (NUREG/CR-4832, Vol. 9). The probability of a significant fire in fire area AC (FIRE-AC) can be represented by:

$$\text{FIRE-AC} = \lambda_{\text{AUX}} \, f_{\text{AAC}} \, Q \, f_{\text{AC}} \, f_{\text{s}} \qquad \text{(D-2)}$$

where

λ_{AUX} = auxiliary building fire frequency

f_{AAC} = area ratio of fire area AC to that of the auxiliary building

Q = probability that the fire will not be manually suppressed before the critical fire-induced damage occurs

f_{AC} = area ratio within fire area AC where a significant fire can damage the critical components

f_{s} = severity ratio for a significant fire

The LaSalle fire modeling has determined that a small fire anywhere in fire area AC can cause the rapid formation of a hot gas layer that fails all critical cabling. Therefore, the room-specific area term (f_{AC}) and the severity ratio (f_{s}) are both 1.0.

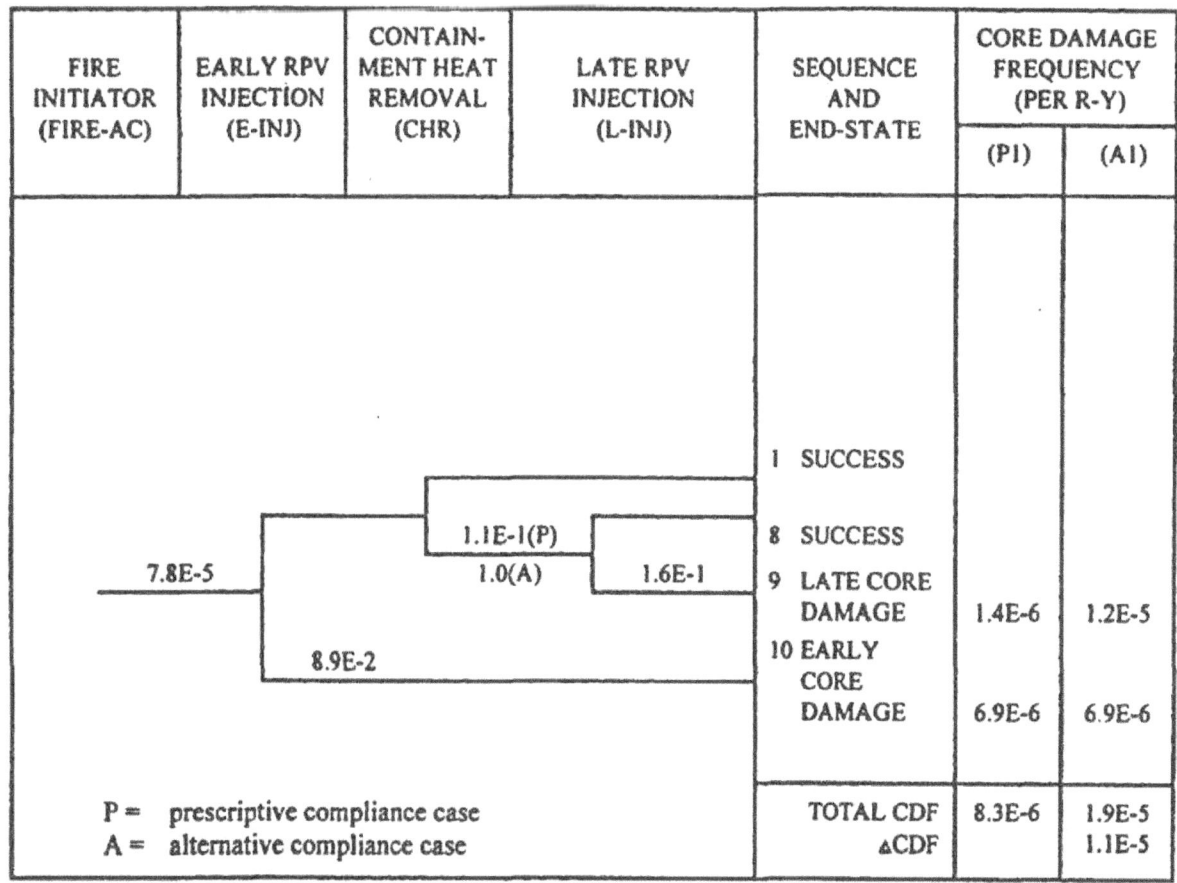

FIRE INITIATOR (FIRE-AC)	EARLY RPV INJECTION (E-INJ)	CONTAIN-MENT HEAT REMOVAL (CHR)	LATE RPV INJECTION (L-INJ)	SEQUENCE AND END-STATE	CORE DAMAGE FREQUENCY (PER R-Y)	
					(P1)	(A1)
		1.1E-1(P) 1.0(A)	1.6E-1	1 SUCCESS		
7.8E-5				8 SUCCESS		
				9 LATE CORE DAMAGE	1.4E-6	1.2E-5
	8.9E-2			10 EARLY CORE DAMAGE	6.9E-6	6.9E-6
P = prescriptive compliance case A = alternative compliance case				TOTAL CDF ΔCDF	8.3E-6	1.9E-5 1.1E-5

Figure D.13
72-Hour Case Study—Quantified Event Tree for Case 1

Similarly, very little credit can be taken for manual fire suppression activities (Q = 0.99) because of the comparatively short time before critical damage occurs. Table D.5 presents the best-estimate values of all terms in Equation D-2 for fire area AC, as well as their associated distributions. Therefore: FIRE-AC = λ_{AUX} f_{AAC} Q f_{AC} f_s = 7.8E-5 per reactor-year

Early RPV Injection (E-INJ)

Early RPV injection is a functional event that consists of systems and combinations of systems that can satisfy immediate and longer term core makeup requirements. For the purposes of this case study, early RPV injection has been simplified by crediting the high-pressure core spray (HPCS) system. The Integrated Reliability and Risk Analysis System (IRRAS) model of the LaSalle Unit 2 PRA (NUREG/CR-5813) is used to estimate the HPCS system unavailability. The

logic model and the failure data in the IRRAS model remain the same. The failure probability of early RPV injection is: E-INJ = 8.9E-2. This value is applicable to both the prescriptive compliance and the alternative case.

Containment Heat Removal (CHR)

Containment heat removal is also a functional event that could credit different systems. For this case study, this top event is approximated by one train of the suppression pool cooling mode of RHR, as modeled in the IRRAS version of the LaSalle PRA.

The prescriptive compliance case assumes that one train of RHR is removed from fire area AC or otherwise protected. Therefore, a failure of the CHR function requires additional RHR random failures.

Table D.5 Distributions of Terms for Core-Damage Equation for Fire Area AC

Factor	Distribution	Best Estimate	Lower Bound	Upper Bound
λ_{AUX}	Gamma	0.049	8.50E-3	0.12
f_{AAC}	Maximum entropy	1.60E-3	3.20E-4	8.00E-3
Q	Maximum entropy	0.99	0.46	1.0
f_{AC}	-	1.0	-	-
f_s	-	1.0	-	-

Source: LaSalle fire PRA (NUREG/CR-4832, Vol. 9).

The estimated unavailability is $CHR_{PI} = 1.1E-1$. The alternate case does not protect the RHR system. All containment heat removal is assumed lost due to the fire, and $CHR_{AI} = 1.0$.

Late RPV Injection, (L-INJ)

The failure of late RPV injection is due to the severe environment in the reactor building after containment failure. Although other systems, such as the control rod drive hydraulic system, may be available, consistent with the LaSalle PRA the assessment of RPV injection after containment failure conservatively considers only the HPCS system. This assumption will conservatively accentuate the importance of the 72-hour requirement in the analysis results. This failure estimate is derived from the IRRAS model of the LaSalle PRA: $L-INJ = 1.6E-1$.

This injection unavailability estimate is applicable to both the prescriptive and the alternative cases.

Figure D.13 provides the CDF for a significant fire in room AC using the modeling assumptions of the LaSalle PRA.

The prescriptive compliance case assumes the B train of RHR is isolated from the effects of the fire. Both of the contributing sequences require additional random (non-fire) failures to reach core damage. This results in a CDF of 8.3E-6 per reactor-year.*

For room AC the alternative case assumes that all decay heat removal is lost due to fire damage. Furthermore, no recovery actions are credited. The result is a CDF of 1.9E-5 which is dominated by the late core-damage sequence number 9. The ΔCDF is a significant 1.1E-5 per reactor-year.

In order to minimize the effects of modeling assumptions on the ΔCDF, it is important to use the same level of resolution to model the prescriptive and alternative approaches. For example, if the alternative approach credited operation action, but the prescriptive case did not, a minimal ΔCDF could be developed. In that case, modeling disparities could mask the true impact of the alternative approach.

In the typical PRA analysis, the sequences or areas that are not major contributors to core damage are generally not modeled in detail. Conservative assumptions are used to allow analytical resources to be dedicated to the more detailed modeling associated with the dominant accident sequences. Case 1 demonstrates that non-dominant sequences in a fire PRA may not be modeled in sufficient detail to permit their use in a realistic assessment of the increase in core-damage frequency associated with an alternative approach. The next section uses a more realistic, more detailed model that examines operator recovery of the containment heat removal function in the 27-hour time period preceding containment failure.

---

* The simplifying assumptions used herein result in a CDF contribution that is approximately one order of magnitude higher than the LaSalle fire PRA analysis of area AC.

Case 2—Refined Modeling Assumptions

As stated before, the LaSalle PRA used conservative modeling assumptions for the non-dominant contributors to the fire-induced CDF. Therefore, the Case 1 ΔCDF is not realistic. The time available (~27 hours) before containment failure allows ample opportunity for the recovery of the containment heat removal function. It is appropriate to examine these recovery efforts by revising the Case 1 event tree.

Case 2 will model the recovery of the power conversion system when containment heat removal is unavailable because of the postulated fire (alterative case) or because of random failures (prescriptive compliance case). Given successful operation of the power conversion system (PCS) for 200 hours, this case also models RHR repair to permit cold shutdown.

From a PRA perspective, Case 2 presents two modeling challenges. Like Case 1, the successful end-states include both stable and cold-shutdown configurations. However, Case 2 considers much longer mission times, based on plant-specific and accident-sequence considerations. One successful end-state is cold shutdown after the repair of one RHR train. This process is estimated to require 200 hours to reach cold shutdown. An alternative success path considers longer term operation of the PCS, resulting in stable shutdown at 400 hours.

The typical PRA must be reevaluated and extended to accurately capture potential systems interactions and important operator actions. This case uses simplifying assumptions and focuses on the long-term PCS operation. A plant-specific analysis is necessary to examine plant capability, system interlocks, procedures, and operator actions.

Second, many PRA models can be expected to place an emphasis on operator action. In this instance, the operator actions to reestablish the condenser and to recover one train of RHR are critical issues. Although Case 2 examines these actions for both the prescriptive compliance and alternative approaches, there can be differing failure estimates, depending on the context. To

accurately estimate the likelihood of success and to minimize uncertainty, detailed plant-specific human reliability analyses are required; however, current state of the art in HRA techniques may limit such analyses. For illustrative purposes, conservative failure estimates were used for these restoration actions.

This case examines the alternative of reestablishing the condenser for long-term decay heat removal to allow sufficient time for the repair of one train of RHR shutdown cooling. In accordance with the definition of stable shutdown, long-term operation of the PCS or continued RPV injection after containment failure are also considered successes.

The accident sequences for Case 2 are presented in Figure D.14. The higher level of modeling resolution results in 10 sequences. Sequence 1 represents successful early RPV injection (E-INJ) and successful containment heat removal (CHR) after a fire in area AC. It is the same as Sequence 1 of Case 1. Sequence 10 describes the near-term failure of RPV injection. It is an early core-damage sequence and is also the same as its counterpart in Case 1.

Unlike the previous case, given a CHR failure, this event tree models the reopening of the MSIVs or main steam line drain valves (REC-PCS) to recover the containment heat removal function. A failure implies ultimate containment failure. Sequences 8 and 9 are conceptually similar to those described in Case 1. Sequence 8 evaluates continued RPV injection despite the harsh environment in the reactor building caused by containment failure. Sequence 9 results in late core damage because of an environmentally induced failure of RPV injection.

Given successful PCS recovery, the tree examines the operation of PCS for 200 hours (PCS-200H). If this top event is not successful, the harsh environment due to containment overpressurization failure again challenges RPV injection. Sequence 6 assumes injection continues. Sequence 7 represents late core damage due to late RPV injection failure.

The successful operation of PCS for a mission

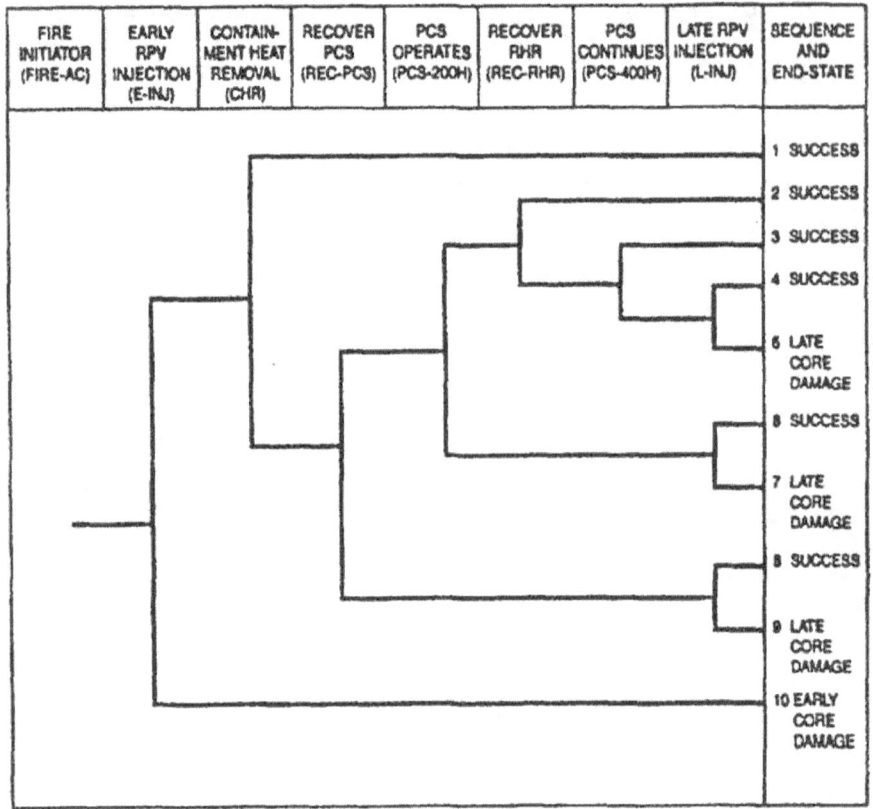

FIRE INITIATOR (FIRE-AC)	EARLY RPV INJECTION (E-INJ)	CONTAIN-MENT HEAT REMOVAL (CHR)	RECOVER PCS (REC-PCS)	PCS OPERATES (PCS-200H)	RECOVER RHR (REC-RHR)	PCS CONTINUES (PCS-400H)	LATE RPV INJECTION (L-INJ)	SEQUENCE AND END-STATE
								1 SUCCESS
								2 SUCCESS
								3 SUCCESS
								4 SUCCESS
								5 LATE CORE DAMAGE
								6 SUCCESS
								7 LATE CORE DAMAGE
								8 SUCCESS
								9 LATE CORE DAMAGE
								10 EARLY CORE DAMAGE

Figure D.14
72-Hour Case Study—Event Tree for Case 2

time of 200 hours will allow one train of RHR to be repaired (top event REC-RHR). Successful repair and operation of RHR will allow cold shutdown to be reached (Sequence 2). If RHR cannot be repaired, continued PCS operation is examined.* Sequence 3 represents stable shutdown using the PCS in lieu of RHR. If PCS fails during this extended mission time, continued RPV injection after containment failure is again modeled as Sequences 4 and 5.

The quantified event tree for Case 2 is presented as Figure D.15. The top events are discussed below.

Initiator (FIRE-AC)

The LaSalle fire modeling of area AC has determined that a small fire can cause the rapid

formation of a hot gas layer that can fail all critical cabling. As before (see Equation D-2), FIRE-AC = λ_{AUX} f_{AAC} Q f_{AC} f_{S}.

Since the geometry of fire area AC, the time to damage the critical cables, and the auxiliary building fire frequency remain unchanged, the values of Table D.6 are appropriate and FIRE-AC = 7.8E-5 per reactor-year.

Early RPV Injection (E-INJ)

The early RPV injection top event is unchanged from Case 1. The failure probability of early RPV injection is E-INJ = 8.9E-2.

Containment Heat Removal (CHR)

The CHR functional event is identical to that used in Case 1. The estimated unavailability for the prescriptive approach is CHR_{P2} = 1.1E-1.

---

*A mission time of 400 hours is arbitrarily assumed.

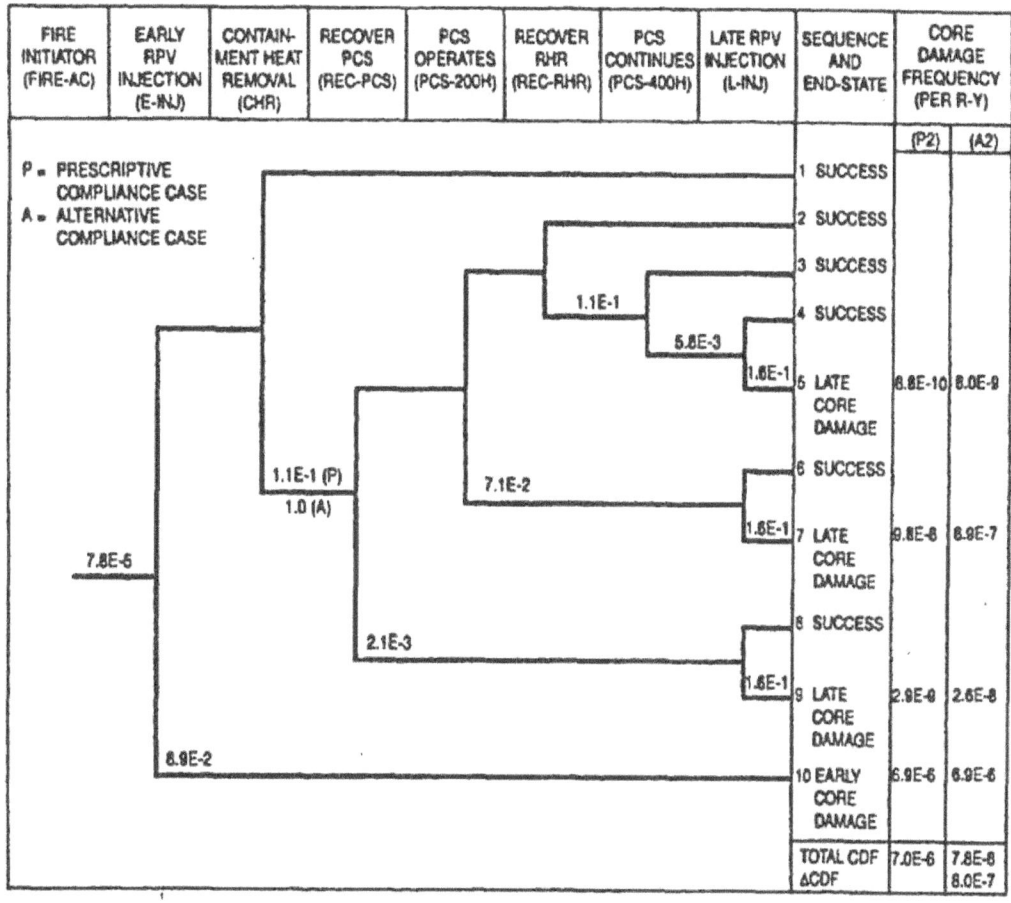

FIRE INITIATOR (FIRE-AC)	EARLY RPV INJECTION (E-INJ)	CONTAIN-MENT HEAT REMOVAL (CHR)	RECOVER PCS (REC-PCS)	PCS OPERATES (PCS-200H)	RECOVER RHR (REC-RHR)	PCS CONTINUES (PCS-400H)	LATE RPV INJECTION (L-INJ)	SEQUENCE AND END-STATE	CORE DAMAGE FREQUENCY (PER R-Y)	
									(P2)	(A2)
								1 SUCCESS		
								2 SUCCESS		
								3 SUCCESS		
				1.1E-1				4 SUCCESS		
					5.8E-3		1.8E-1	5 LATE CORE DAMAGE	8.8E-10	8.0E-9
		1.1E-1 (P)		7.1E-2				6 SUCCESS		
		1.0 (A)					1.6E-1	7 LATE CORE DAMAGE	9.8E-8	8.9E-7
7.8E-5			2.1E-3					8 SUCCESS		
							1.6E-1	9 LATE CORE DAMAGE	2.9E-8	2.6E-8
	8.9E-2							10 EARLY CORE DAMAGE	6.9E-6	6.9E-6
								TOTAL CDF ΔCDF	7.0E-6	7.8E-6 8.0E-7

Figure D.15
72-Hour Case Study—Quantified Event Tree for Case 2

Since the alternative approach does not protect the RHR system, CHR$_{A2}$ = 1.0, as before.

Failure To Recover the PCS (REC-PCS)

PCS recovery is necessary to ensure long-term decay heat removal. A plant-specific human reliability analysis is required to estimate the failure probability of this recovery action, but this kind of analysis is outside the scope of this case study. A value of 2.1E-3 has been adopted from the LaSalle PRA. It represents the failure to manually open the main steam line drain valves to depressurize the RPV: REC-PCS = 2.1E-3

Failure of the PCS To Operate for 200 Hours (PCS-200H)

The IRRAS model of the LaSalle Unit 2 PRA

(NUREG/CR-5813) is used to evaluate the PCS. The logic model and the failure data in the IRRAS model remain the same. The failure of the PCS to operate for a 200-hour mission time is PCS-200H = 7.1E-2.

Failure To Recover One Train of RHR (REC-RHR)

Normally, recovery efforts are required to be completed in shorter times than the 200 hours assumed here. When a comparatively short amount of time is available for recovery actions, human error generally dominates and any hardware failures that could prevent the recovery are inconsequential. In our case, however, the 200-hour time window results in a low estimate of the human error rate. The failure to recover RHR is dominated by hardware failures and is approximated by the CHR top event, i.e., the unavailability of a single train of suppression pool cooling. No additional repairs are assumed. Therefore, the failure to recover RHR is 1.1E-1 for both the prescriptive compliance and the alternative approaches, REC-RHR = 1.1E-1.

Failure of the PCS To Continue To Operate After 200 Hours (PCS-400H)

If RHR is not recovered and cold shutdown cannot be reached, the continued operation of the PCS to maintain stable shutdown is also credited. For the purposes of this study, this event considered only the failure of a circulating water pump, the failure of a mechanical vacuum pump, and the potential for the loss of offsite power during the additional mission time of 200 hours. A plant-specific analysis would include an analysis of plant capability, system interlocks, procedures, and operator actions, PCS-400H = 5.8 E-3.

Late RPV Injection (L-INJ)

The failure of late RPV injection (HPCS)[*] is due to the severe environment in the reactor building after containment failure. This failure estimate is unchanged from Case 1,
L-INJ = 1.6E-1

The evaluation of all the headings of the event tree of Figure D.14 is presented in Figure D.15, and the four sequences leading to core damage are quantified for both the prescriptive and the alternative approaches. The final result is given at the bottom of Figure D.15; it is ΔCDF = 8.0E-7.

D.3.3 Uncertainty Analysis

Thus far, this case study has used mean values to evaluate the ΔCDF of alternative approaches to prescriptive regulation. However, point estimates do not reflect the inherent variability in the data and modeling uncertainties. One of the chief criticisms of a point estimate model is that the

result does not provide an understanding of the range of values that the outcome is likely to assume. That requires an uncertainty analysis.

This section will summarize an uncertainty evaluation that was performed for the 72-hour case study. A ΔCDF distribution as a function of cumulative probability is developed for each case study. The uncertainty ranges for the two cases are compared. In lieu of the point estimate, a conservative percentile value of the ΔCDF is chosen to reflect the various sources of uncertainty.

The uncertainty analysis for this case study is relatively straightforward, primarily because no credit is taken for fire modeling.[**] Only PRA techniques were used to compare the prescriptive compliance and the alternative approaches. Risk assessments such as LaSalle fire PRA routinely include formal uncertainty analyses, and the techniques are well established. The uncertainty information for this case study was generally adopted from the LaSalle PRA (NUREG/CR-4832). Several volumes of this analysis are devoted to parameter estimation, the human reliability evaluation, and the uncertainty analysis.

The LaSalle PRA calculated an uncertainty importance for each of the dominant sequences. For a fire in room AC, the percent reduction in the uncertainty of log risk is dominated (~88 percent) by the uncertainty associated with equipment survivability after primary containment overpressurization failure. Parameters that are related to fire initiation and propagation are relatively small contributors to the uncertainty importance. In general, most of the parameter distributions in the LaSalle PRA are assumed to be log normal, although several basic events used other distributions or user-specified distributions. For the purposes of this case study, the latter events are approximated by the lognormal distributions so that the IRRAS code could be used to calculate the uncertainty range for each

---

[*] Consistent with the LaSalle PRA, the failure of RPV injection after containment failure conservatively considers only the high-pressure core spray system.

[**] This case study adopts the LaSalle PRA assumption that a small fire anywhere in room AC will cause the rapid formation of a hot gas layer that causes all critical cabling to fail.

case.

The IRRAS code is used to calculate the CDF uncertainty for this case study. One thousand CDF samples are generated for each of the sequences presented in Figures D.13 and D.15. A FORTRAN program and a spreadsheet are used to combine each sample and generate 1,000 ΔCDF values for each case.

The distribution of the ΔCDF using conservative modeling assumptions (Case 1) is presented as Figure D.16. Unlike the point estimate of 1.1E-5 developed earlier, this distribution provides a feel for how much the ΔCDF can vary. One way to account for the distribution is to specify a confidence level instead of a point estimate. For example, a 90-percent confidence criterion results in a ΔCDF value of 1.5E-5.

Figure D.17 presents the cumulative probability distribution for the ΔCDF using refined modeling assumptions (Case 2). The 90th percentile ΔCDF for Case 2 is about 1.1E-6.

Normally, cases that use more detailed modeling and that take credit for additional human actions have greater uncertainty bands when compared to simpler, more conservative models. However, as shown in Figures D.16 and D.17, the uncertainty bands between the 10-percent and the 90-percent confidence limits are roughly comparable. This is attributable to the dominance of the uncertainty associated with continued injection after containment failure. This tends to mask the recovery uncertainty associated with Case 2.

In general, we would expect cases that feature higher levels of modeling resolution, particularly those that credit human actions, to have greater uncertainty bands. However, the use of a conservative confidence limit will capture this increased uncertainty.

D.3.4 Summary

This case study examines the safety impact of alternative approaches to the 72-hour criterion to reach cold shutdown. Several key considerations are summarized below:

The fire regulations in 10 CFR Part 50 are designed to protect the health and safety of the public by helping to assure that safe cold shutdown can be achieved. PRAs also examine public risk, but different assumptions are used. Postulated failures are not subject to regulatory constraints, i.e., "a single active failure." Success is also defined differently. The typical Level 1 100-percent- power PRA considers various transitional end-states to be successes, even though cold shutdown has not been reached.[*] These stable shutdown end-states do not pose additional challenges to key critical safety functions and the core is expected to remain intact. From a PRA perspective, these sequences are not dominant and additional modeling will not significantly change the CDF or the analytical insights.

This basic difference between the regulatory and the PRA definitions of success needs to be addressed for risk-informed and performance-based regulation. Is it necessary to specify cold shutdown as the only successful end-state? On the other hand, is it appropriate from a regulatory perspective to allow the failure of a major fission product barrier such as the containment or the fuel rods? As part of the PRA process, screening analyses are generally performed to identify the major contributors to risk (or CDF for the Level 1 PRA). Dominant initiators, systems, and sequences are identified for more detailed evaluation. Dominant sequences may utilize several detailed system fault trees for a single top event; human errors might be quantified using a simulator; and recovery actions are developed, quantified, and credited where appropriate. Non-dominant sequences generally are quantified using the conservative screening assumptions. For example, Case 1 of this case study considered only HPCS for RPV injection, and recovery actions were not credited.

As illustrated by this case study, alternate approaches can be expected to require reexamination of non-dominant sequences. Case

---

[*] A Level 2 PRA might define success as core damage, but no release occurs because the containment remains intact.

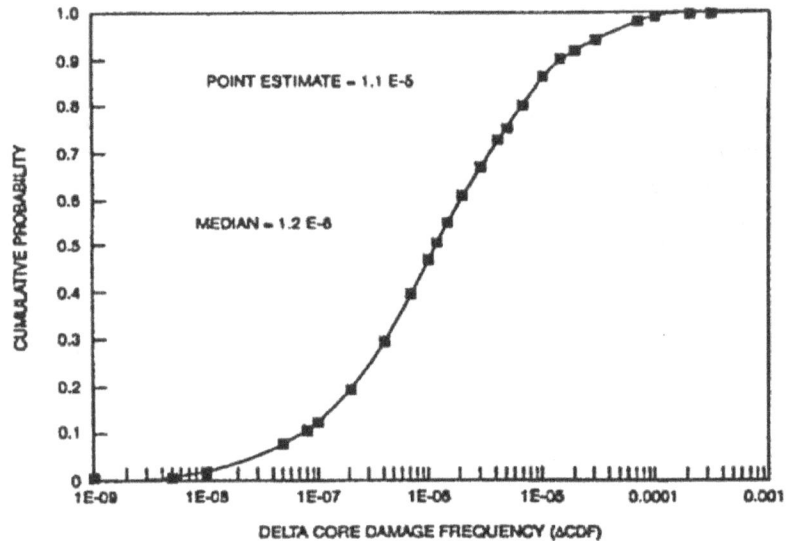

Figure D.16
Cumulative Probability Function for Case 1

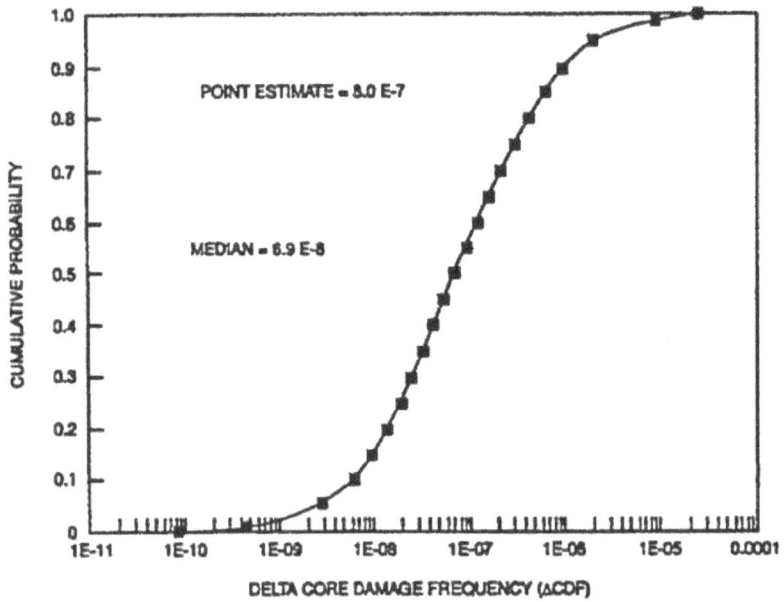

Figure D.17
Cumulative Probability Function for Case 2

2 uses a finer level of modeling resolution to credit certain operator recovery actions, events that are commonly modeled in current PRAs. However, long- term operation of the PCS is also modeled in Case 2. To the best of our knowledge, this has not been considered elsewhere.* Finer levels of modeling resolution, crediting increasingly complex operator actions and unusual system configurations, could have been employed herein. The consistent application of risk-informed performance-based initiatives will require a consensus on the level of modeling resolution that is appropriate.

Section D.3.3 presents the uncertainty analyses for this case study. The results are presented as probability distributions of the ΔCDF that help the reader to assess the variability of the input. A 90-percent confidence limit was chosen for illustrative purposes.

This case study was particularly suitable for uncertainty analysis because it did not credit any fire modeling. Unlike fire modeling, uncertainty analysis techniques for PRAs for internal event sequences are well established.

This analysis is dominated by a single event,

---

* Probably because it would be considered stable shutdown as discussed above.

"continued RPV injection after containment failure." As a "rare" event, the uncertainty band was established primarily by expert opinion.**

Within the regulatory context, the reliance on expert opinion for events that dominate uncertainty should be assessed. This is not an intractable problem; a suitable confidence band could be specified. For example, in this case study, a Δ core damage criterion of 1E-5 could be satisfied at 99-percent confidence level for case 2. This could be construed as a probabilistic safety margin. Alternatively, different modeling assumptions could be employed to avoid the dominant source of uncertainty.

This case study uses ΔCDF as a tool toward evaluating the safety equivalence of an alternative approach to a prescriptive requirement. Several issues have been raised for further evaluation. These issues notwithstanding, a probabilistic approach provides a consistent framework in which to identify key issues, examine sensitivities, and evaluate the safety equivalence of an alternative approach to a prescriptive requirement.

** The experts provided input on containment failure locations and sizes. This information was used to calculate time-temperature profiles for various reactor building locations.

NRC FORM 335
(2-89)
NRCM 1102.
3201, 3202

U.S. NUCLEAR REGULATORY COMMISSION

BIBLIOGRAPHIC DATA SHEET

(See instructions on the reverse)

1. REPORT NUMBER
(Assigned by NRC, Add Vol., Supp., Rev., and Addendum Numbers, if any.)

NUREG-1521

2. TITLE AND SUBTITLE

Technical Review of Risk-Informed, Performance-Based Methods for Nuclear Power Plant Fire Protection Analyses

3. DATE REPORT PUBLISHED

MONTH	YEAR
July	1998

4. FIN OR GRANT NUMBER

5. AUTHOR(S)

M. K. Dey, NRC
M. A. Azarm, R. Travis, G. Martinez-Buridi, BNL; R. Levine, NIST

6. TYPE OF REPORT

Draft for Comment

7. PERIOD COVERED (Inclusive Dates)

N/A

8. PERFORMING ORGANIZATION - NAME AND ADDRESS (If NRC, provide Division, Office or Region, U.S. Nuclear Regulatory Commission, and mailing address; if contractor, provide name and mailing address.)

Division of Systems Technology
Office of Nuclear Regulatory Research
U.S. Nuclear Regulatory Commission
Washington, DC 20555

9. SPONSORING ORGANIZATION - NAME AND ADDRESS (If NRC, type "Same as above"; if contractor, provide NRC Division, Office or Region, U.S. Nuclear Regulatory Commission, and mailing address.)

Same As Above

10. SUPPLEMENTARY NOTES

11. ABSTRACT (200 words or less)

The Nuclear Regulatory Commission (NRC) has instituted an initiative for regulatory improvement to focus licensee and NRC resources on risk-significant activities, and decrease the prescriptiveness of its regulations. The NRC has identified risk-informed methods utilizing insights from probabilistic risk analysis (PRA) as a major tool for achieving its goal for regulatory focus. Fire protection requirements has been identified as a regulatory area in which NRC will pursue regulatory improvement. This report presents a technical review and analysis of alternative risk-informed, performance-based methods to those in current prescriptive fire protection requirements or guidance that could allow cost-effective methods for implementation of safety objectives, focus licensee efforts, and achieve greater efficiency in the use of resources for plant safety. A technical analysis of the usefulness of the results and insights derived from these methods (including accounting for the uncertainties in the results) in improving regulatory decision making is presented.

12. KEY WORDS/DESCRIPTORS (List words or phrases that will assist researchers in locating the report.)

Probabilistic Risk Analysis
Fire Model
Fire Protection
Performance-Based Regulation
Risk-Informed Regulation
Regulatory Review and Improvement
Nuclear Power Plant

13. AVAILABILITY STATEMENT

unlimited

14. SECURITY CLASSIFICATION

(This Page)

unclassified

(This Report)

unclassified

15. NUMBER OF PAGES

16. PRICE

This form was electronically produced by Elite Federal Forms, I